Global Environmentalism
and Local Politics

SUNY series in Global Environmental Policy

Uday Desai, editor

Global Environmentalism and Local Politics

Transnational Advocacy Networks in Brazil, Ecuador, and India

Maria Guadalupe Moog Rodrigues

STATE UNIVERSITY OF NEW YORK PRESS

Published by
State University of New York Press, Albany

© 2004 State University of New York

All rights reserved

Printed in the United States of America

No part of this book may be used or reproduced in any manner whatsoever without written permission. No part of this book may be stored in a retrieval system or transmitted in any form or by any means including electronic, electrostatic, magnetic tape, mechanical, photocopying, recording, or otherwise without the prior permission in writing of the publisher.

For information, contact State University of New York Press, Albany, NY www.sunypress.edu

Production by Kelli Williams
Marketing by Michael Campochiaro

Library of Congress Cataloging-in-Publication Data

Rodrigues, Maria Guadalupe Moog.
  Global environmentalism and local politics : transnational advocacy networks in Brazil, Ecuador, and India / Maria Guadalupe Moog Rodrigues.
     p. cm. — (SUNY series in global environmental policy)
  Includes bibliographical references and index.
  ISBN 0-7914-5877-6 (hc : alk. paper) — ISBN 0-7914-5878-4 (pbk. : alk. paper)
    1. Sustainable development—Developing countries—Citizen participation—Case studies. Green movement—Developing countries—Case studies. 3. Non-governmental organizations—Developing countries—Case studies. 4. Globalization—Case studies. I. Title. II. Series.

HC59.72.E5R63 2004
333.72'098—dc22

2003061096

10 9 8 7 6 5 4 3 2 1

CONTENTS

Abbreviations and Acronyms	vii
Preface	xi
Acknowledgments	xiii
1. Introduction	1
Transnational Advocacy Networks, Civil Society, and the Environment: Defining Concepts	5
Key Questions	12
Research Method and Organization of the Book	15
2. The Dilemma of Amazonian Development and Its Impact on Rondônia	19
Amazonian Development and Military Rule (1964–1984)	21
Environmental and Development Policies for Amazonia in Democratic Brazil	26
3. Urgent Action! Transnational Mobilization against Disaster in Rondônia	33
The Beginning	33
The Rondônia Network Gains Momentum	36
Transnational Activism and Domestic Constraints— The Rondônia Network within Brazil	40
Taking Stock of the Rondônia Network in the 1980s	43
4. "Localizing" Transnational Activism—Success and Failure	49
The Issue of Transparency: Environmentally Sustainable Development to Whom? By Whom?	51
Building Capacity for Participation: Searching for the Right Recipe in Rondônia	55

The Rondônia Network's Legitimacy Crisis of 1994	60
Taking Stock of the Rondônia Network in the Early 1990s	64

5. Listening to The Grassroots—
   The Rondônia Network and Local Politics   67

   The Rondônia Network Takes Planafloro to the Inspection Panel   68
   Evaluating and Restructuring Planafloro: A New Meaning for Environmentally Sustainable Development?   77
   Taking Stock of the Rondônia Network at the Turn of the Millennium   87

6. Environmental Activism beyond Brazil I—
   The Struggle against Oil Exploitation in Ecuador   93

   Fighting Oil Exploitation in Ecuador's Oriente: *Background and Network Origins*   94
   The Anti-Oil Network and the Campaigns against Texaco and ARCO   102
   Taking Stock of Ecuador's Anti-Oil Network   110

7. Environmental Activism beyond Brazil II—
   The Struggle against Large Dams in India   115

   The Struggle against the Sardar Sarovar Dam—Background and Network Origins   116
   The Narmada Network Strategies and Their Local Impacts   121
   Taking Stock of the Narmada Network   128

8. Conclusions   135

   Local Empowerment and Local Results   136
   Lessons and Recommendations   145
   Final Thoughts   148

Notes   151

References   181

Index   191

ABBREVIATIONS AND ACRONYMS

ASODIRA	Association for Indigenous Development
ASP	Amerindian Special Project
CD	Conselho Deliberativo (Deliberative Council)
CESR	Center for Economic and Social Rights
CIEL	Center for International Environmental Law
CIMI	Conselho Indígena Missionário (Indigenous Peoples' Missionary Council)
CNS	Conselho Nacional dos Seringueiros (National Council of Rubber Tappers)
COICA	Coordinating Body of Indigenous Organizations of the Amazon Basin
CONAIE	National Confederation of Indigenous Nationalities of Ecuador
CONFENIAE	Confederation of Indian Nations of Ecuadorian Amazon
CORDAVI	Corporación de la Defensa de la Vida (Corporation for the Defense of Life)
CPT	Comissão Pastoral da Terra (Pastoral Commission on Land)
CUT	Central Única dos Trabalhadores (Central Labor Union)
CUNPIR	Coordenação da União das Nações e Póvos Indígenas de Rondônia (Coordination of the Union of Indigenous Peoples and Nations of Rondônia)
DNPM	Departamento Nacional de Prospecção Mineral (National Department of Mineral Research)

ECOPORE	Ação Ecológica Vale do Guaporé (Ecological Action for the Guaporé Valley)
EDF	Environmental Defense Fund, since 2000 Environmental Defense
EIA	Environmental Impact Assessments
FCUNAE	Quichua Federation of Indigenous Nations of Ecuador
FETAGRO	Federação dos Trabalhadores Agrícolas de Rondônia (Federation of Agricultural Workers of Rondônia)
FIERO	Federação das Indústrias do Estado de Rondônia (Federation of Rondonian Industries)
FIPE	Fundação Instituto de Pesquisas Econômicas (Institute of Economic Research Foundation)
FIPSE	Independent Federation of the Shuar People of Ecuador
FoE	Friends of the Earth
FUNAI	Fundação Nacional do Indio (Brazilian Indian Agency)
IAMA	Instituto de Antropologia e Meio Ambiente (Institute of Anthropology and the Environment)
IBAMA	Instituto Brasileiro do Meio Ambiente e Recursos Naturais Renováveis (Brazilian Institute for the Environment and Renewable Natural Resources)
IBDF	Instituto Brasileiro de Desenvolvimento Florestal (Brazilian Institute for Forestry Development)
IDA	International Development Association
IEA	Instituto de Estudos Amazônicos e Ambientais (Institute for Amazonian and Environmental Studies)
IGO	International Governmental Organization
IRN	International Rivers Network
INCRA	Instituto Nacional de Colonização e Reforma Agrária (National Institute for Colonization and Agrarian Reform)
INDIA	Instituto de Pesquisa em Defesa da Identidade Amazônica (Institute for Research in Defense of Amazonia's Identity)

IPHAE	Instituto de Pré-História, Antropologia e Ecologia (Institute for Pre-History, Anthropology, and Ecology)
MDB	Multilateral Development Bank
MST	Movimento dos Trabalhadores Rurais Sem Terra (Landless Peoples' Movement)
NBA	Narmada Bachao Andolan
NGO	Non-Governmental Organization
NOVIB	Dutch Organization for International Development Cooperation
NRDC	Natural Resources Defense Council
NWF	National Wildlife Federation
OINCE	Organization of the Cofan Indigenous People of Ecuador
OISE	Organization of the Secoya Indigenous People of Ecuador
ONISE	Organization of the Siona Indigenous People of Ecuador
OPIP	Organization of Indigenous Peoples of Pastaza
OSR	Organização dos Seringueiros de Rondônia (Organization of Rondonian Rubber Tappers)
PACA	Proteção Ambiental Cacoalense (Environmental Protection of the Cacoal Region)
PAIC	Programa de Apoio a Iniciativas Comunitárias (Program for the Support of Community Initiatives)
PIN	Programa de Integração Nacional (National Integration Program)
PLANAFLORO	Plano Agropecuário e Florestal de Rondônia (Rondônia Agro-livestock and Forestry Plan). The project's official name at The World Bank is the Rondônia Natural Resources Management Project.
PNMA	Programa Nacional de Meio Ambiente (National Program for the Environment)
POLONOROESTE	Programa de Desenvolvimento Integrado do Noroeste do Brazil (Northwest Brazil Integrated Development Program)
RAN	Rainforest Action Network

SAE	Secretaria de Assuntos Estratégicos (Secretariat of Strategic Affairs)
SIVAM	Sistema de Vigilância da Amazônia (System for the Surveillance of Amazonia)
SSP	Sardar Sarovar Project
UNCED	United Nations Conference on Environment and Development
UNEP	United Nations Environmental Program
UNI	União Nacional Indígena (National Indigenous Union, Rondônia chapter)
WWF	World Wildlife Fund

PREFACE

Twelve years ago, as a masters student in International Relations, I heard the following statement from one of the most prominent professors in the field: "While your research project is well drafted, I strongly recommend that you choose a different topic. Non-governmental organizations (NGOs) simply play no role in international politics. In the end, states are all that matter." Today I am glad I did not follow his advice. NGOs, and in particular the networks that they have built across nations, bringing together actors of different natures and nationalities (individuals, research and advocacy organizations, social movements, among others) do matter. They have influenced politics and policymaking at the international, national, and local levels. They have affected the way states conduct business in the areas of human rights, the environment, development aid, and women's rights. And if there are still those who think that NGOs and other actors in the global civil society may only affect "soft issues," they should be reminded of the events that surrounded the meeting of the World Trade Organization in Seattle in 1999.

This book is the result of ten years of academic, professional, and personal interest in understanding how the common citizen may and does participate in international politics. While I believe in democratic representation, I do not think that the possibility of electing a representative should forfeit the right—and the duty—of concerned individuals to have a say in issues that matter to them. In an increasingly interdependent world, such issues have multiplied. Transnational activism is an important instrument in the process of expanding the scope of democratic participation in international politics. Transnational advocacy networks, in particular, have been powerful instruments in creating opportunities for the exercise of one's "global citizenship."

The possibility of affecting processes and decisions that are global in scope is particularly relevant in the case of the environment. Environmental problems in one end of the earth may be relevant to an individual residing at the opposite end, not simply as a matter of principle or ideology. More often then not, such problems gain relevance because they have material consequences—they can cause harm to individuals and/or societies beyond the traditional limits of geography and politics. Environmental problems thus differ in nature from other "global problems." Let us consider, for instance, human rights. One may feel that he or she has the duty to engage in advocacy initiatives to guarantee that the rights of all human beings are protected. Such a feeling is based, exclusively, on one's principles and ideals, since the denial of another's right to physical integrity, for instance, has no consequence for one's own. Differently, if ecosystems are destroyed, CO^2 emissions increase, and water becomes scarce, one may feel not only the *duty* to try to alter such processes, but also the *right*, as his or her own material well-being may be at stake.

A personal interest in the boundaries of the rights and duties of individuals and organizations concerned with environmental problems underlay my research for this book. I was impressed by the intense commitment to environmental and human rights struggles displayed by activists of different nationalities, backgrounds, and spheres of action. The same was true, in many instances, among individuals working in international and domestic development agencies. Witnessing such dedication was the most gratifying aspect of the research. Inevitable frustration derived from the realization that, in the past twenty years, despite the levels of individual and organizational dedication to the promotion of a healthier global environment, great difficulties still remain. This book addresses some of these difficulties; particularly the ones inherent to the process of mobilizing civil society actors on behalf of the environment. It is my hope that its lessons may contribute to the effectiveness of global environmental activism in the coming years.

ACKNOWLEDGMENTS

I owe an enormous debt of gratitutde to the men and women who granted me interviews during my research for this book, sharing with me not only the details of their work, but also their dreams for a healthier and more just world. While I cannot name them all here, I would like to thank Stephen Schwartzman and Barbara Bramble for being such great sources of inspiration when my work for this book was still in its embryonic stage. As my research unfolded, Aurelio Vianna, Brent Millikan, and José Maria dos Santos were unyielding in their support and availability through the years. I also thank Jonathan Fox, my academic mentor and now colleague, and my advisors at Boston University, Betty Zisk and David Mayers, for their guidance and faith in my work. Thanks to Marielle and Paco, for helping me with the research in Ecuador, and to my colleagues at Holy Cross, Aldo, Steve, Vickie, and Judith, for encouraging me to "wrap it up." Fieldwork would not have been possible without a Batchelor-Ford Summer Faculty Fellowship, from the College of the Holy Cross, a grant from the Committee on Fellowships, Research and Publication, also from the College of the Holy Cross, and a grant from the Institute for the Study of World Politics (ISWP). Finally, I am forever thankful to my parents, Aroldo and Anna Maria, for being the best academic cheerleaders one can have, and to my husband Ram, for—literally—everything.

1

Introduction

The organization Friends of the Earth bought our tickets to Washington, D.C. The occasion was the mobilization for the celebration of the fifty years of the multilateral financing institutions and the climax of the international campaign against them, known as "Fifty Years is Enough!" We went there to denounce irregularities with the World Bank–funded Planafloro project in Rondônia. I went, together with Luisinho, the executive secretariat for the NGOs' Forum of Rondônia, and Almir, a Surui Indian. We were to meet the representatives of Friends of the Earth and Oxfam. We lived the biggest adventure of our lives! Nobody was waiting for us in the Washington, D.C., airport. We managed to find the buildings where the meetings were happening, but I do not know how, since the only thing Luisinho could say in English was to ask if anybody spoke either Portuguese or Spanish. . . . When we arrived at the convention area, there were dozens of meetings happening at the same time, in a gigantic convention center. It was a madhouse in there! We were totally lost. We kept walking up and down the corridors and security was starting to ask us questions since we were the only different people around—we were wearing T-shirts that demanded the creation of extractive reserves and Almir was wearing a traditional headband decorated with colorful feathers. It was then that Patricia, from Oxfam, showed up. With her help, everything became easier. She and Smeraldi, from Friends of the Earth, arranged meetings between us and World Bank executive directors, and we were able to gain the support of a group of directors for the cause of protecting Amazonia's environment. (José Maria dos Santos, president of the Organization of Rondonian Rubber Tappers)[1]

The successful epilogue of the "adventure" of the president of the Organization of Rondonian Rubber Tappers (OSR) illustrates the strategies and processes of resource sharing that has characterized modern environmental politics. This type of politics is one example of the new trends in global environmental governance[2] whose understanding has challenged traditional concepts and frameworks of analysis. State-centric perspectives, for instance, even when conceived in terms of interstate cooperation or regimes, have limited explanatory power to assess environmental management initiatives fostered by non-state actors, such as those attempted by the Organization of Rondonian Rubber Tappers, Friends of the Earth, and others.

Fortunately, since the 1990s, approaches to global environmental governance have broadened the scope of analysis to account both for non-state actors involved in environmental politics and for the transnational nature of environmental issues. These approaches have gone beyond the analysis of processes at the level of the nation-state, and looked both "downward," toward forces operating inside states, and "upward," toward the international system and the actors active in it—multilateral organizations, international corporations, international non-governmental organizations (NGOs) and social movements, and the global civil society.

The study of transnational environmental advocacy networks is particularly relevant precisely because the objects of analysis (the networks themselves) operate, simultaneously, at the local, national, and international levels. In addition, they have been responsible for many of the victories of the global environmental movement to date.[3]

The term *transnational advocacy network* has been common currency in international and comparative politics since the publication of the widely acclaimed and award-winning book *Activists Beyond Borders*, by Margaret Keck and Kathryn Sikkink. In it Keck and Sikkink define transnational advocacy networks as networks of "relevant actors working internationally on an issue, who are bound together by shared values, a common discourse, and dense exchange of information and services."[4] Of particular conceptual significance is the authors' justification for choosing "network" over coalition or movement. The choice was determined by the objects of study themselves, namely, individuals and organizations that participate in such initiatives.[5] Be that as it may, the concept of networks has a long tradition both in sociology and social movements theory as well as in international relations. Transnational advocacy networks' organizational flexibility, capacity to produce and disseminate information, and ability to operate across national borders are important assets in international environmental politics.

The literature on transnational advocacy networks evaluates their impact on global environmental management by focusing on two different arenas. On the one hand are the studies that assess impact on the nation-state and on International Governmental Organizations (IGOs). Some authors, for instance, look at the role that networks play in lobbying governmental officials toward the formulation of environmental treaties and domestic policies, and the creation of environmentally related international lines of credit.[6] Others investigate how transnational advocacy networks have affected reform processes within IGOs leading to the formulation of social and environmental guidelines and safeguards procedures.[7]

The other arena of impact of transnational advocacy networks that existing literature addresses are larger collectivities throughout the world (or entities that participate in world civic politics).[8] In this case, analyses focus on the capacity of transnational advocacy networks to influence international public opinion or the electorate in a given country, and on their role in "translating" the different social meanings of particular struggles (for environmental preservation and indigenous rights, for instance) to stakeholders at different levels: local, national, and international.

Whether evaluating transnational environmental advocacy networks in terms of their impact on states and IGOs or on world civic politics, available literature perceives such networks as a constant. A network's strategies, the sociopolitical and economic contexts in which it operates, and the alliances it builds are variables that affect its capacity to influence its "targets." In this book, I propose a third avenue for the evaluation of the impacts of transnational environmental advocacy networks: the investigation of their impact on the level of empowerment of networks' local members, and as a consequence, on local politics. To achieve this goal, one must open the "black box" and look at transnational networks from the inside out. In this sense, the effectiveness of transnational advocacy networks is very much—although not exclusively—a function of their own internal dynamics, such as their internal politics, resources management, and degrees of cohesion and legitimacy.

When looking at such networks from the inside out, one confronts—and questions—two existing assumptions. The first is an obvious one: that participation in transnational advocacy networks empowers local network members.[9] Studies have claimed, for instance, that alliances and coalitions with international human rights and environmental groups have given "voice" and visibility to local grassroots groups, such as indigenous peoples or *campesino* associations, in national and international arenas. As a consequence, these groups' leverage vis-à-vis opposing forces has increased. A second and related assumption in the literature on transnational advocacy networks is that international and domestic non-governmental organizations play the determinant role in a network's effectiveness.[10] Among all network members (local grassroots groups, individual activists, concerned media), international NGOs, and to a lesser extent their domestic counterparts, possess most of a network's resources and are the ones who make them available to less resourceful network members. Their extraordinary institutional flexibility provides crucial mediation between "levels" or arenas of action.

Without disregarding the role of international and domestic NGOs, I argue that the effectiveness of a transnational environmental advocacy network depends, primarily, on the role that local member organizations play in determining the network's goals and strategies. This discussion is futile unless one determines what counts as "effectiveness." Essentially, effectiveness is a function of goals. If the goal of a transnational environmental advocacy network is to change the behavior of states and international organizations, effectiveness means changing such behaviors. If the goal of a network is to engage the world in civic (environmental) politics, then a high degree of engagement determines effectiveness. There are other, more limited, "measures" of effectiveness. Transnational advocacy networks have been relatively effective in making the World Bank more publicly accountable, and have been successful avenues through which civil society groups can influence a powerful development agency.[11] Yet, none of these "measures" of effectiveness address what I consider the ultimate goal of environmental protection initiatives: the protection of the local environment.[12]

In this book, a transnational environmental advocacy network is effective if and when its members succeed in devising and implementing measures that promote local environmental preservation. These processes are heavily dependent upon the nature of a network's local membership base.[13] In turn, the nature of a network's local membership base is shaped by various processes of "localizing" transnational activism. These mechanisms may or may not lead to the empowerment of local network members. Rather than being an inevitable outcome, as it is widely assumed, the notion that local groups are empowered by participating in transnational advocacy networks requires qualification. Important steps in this process are to define empowerment in specific (local) contexts and to distinguish between political and technical empowerment (while remaining mindful of the relations between the two processes).

For the purposes of the analysis presented in this book, local political empowerment is a function of the establishment of institutionalized mechanisms for local groups' participation in environmental and development policymaking (such as an NGOs' forum or umbrella organization, or the election of groups' representatives to local or national decision-making arenas). It is also a function of their capacity to formulate a common local agenda of priority issues related to environmental protection and development, which implies reaching some level of consensus among different groups affected by a given policy or initiative. Finally, political empowerment is a function of the consolidation of local groups' autonomy vis-à-vis their own national and international network partners as well as in relation to other local political forces (the state and local economic elites, for instance). Because I am particularly interested in transnational environmental advocacy networks, local political empowerment correlates to the technical capacity of local members of a transnational environmental advocacy network to promote environmentally sustainable development. Technical empowerment is thus a function of local groups' capacity to mobilize financial resources to attract (and retain) competent cadres and to make their work operational (access to domestic and international traveling, and to information technology infrastructure, for instance), to provide technical training on environmental and participatory issues to new and existing personnel, and to develop permanent mechanisms for information production and information sharing with other network members, their rank and file, governmental agencies, and the media.

By qualifying the potential role of transnational advocacy networks in empowering their local members I avoid the dangers of a circular argument (the effectiveness of environmental advocacy networks is a function of their local membership base, who is empowered by the network). In fact, while there have been many instances in which participation in transnational advocacy networks has contributed to the political empowerment of local groups in the short term, the absence of a corresponding level of technical and material empowerment has undermined these groups' political position in the long term. There is a perverse irony in the fact that, in many instances, transnational advocacy networks create conditions for the political empowerment of local civil society groups, only to see these groups lose ground under the technical and material burden of their own success.

The relationship between the performance (or effectiveness) of transnational advocacy networks and the level and nature of activism of their local membership base is still an underexplored area of research. Improved knowledge on such a relationship may contribute to a better understanding of the links between global and local civil societies, and on how institutions and processes established in one arena affect dynamics in the other.

TRANSNATIONAL ADVOCACY NETWORKS, CIVIL SOCIETY, AND THE ENVIRONMENT: DEFINING CONCEPTS

The study of initiatives that have bound together actors of different natures who operate at several levels (local, national, and international) has picked up speed since the late 1980s. The fall of the Berlin Wall symbolized the end of a bipolar world and consolidated trends toward interdependence and cooperation in the international system. These trends did not affect nation-states alone. In fact, they became increasingly evident in the dealings of non-state actors such as private corporations, multilateral agencies, and non-governmental organizations of various kinds (churches, trade associations, environmental and human rights organizations, among others).[14] Over time, transnational advocacy networks have become one of the most active sets of actors in certain areas of international politics, such as human rights, the environment, health, and women's issues.[15] While I have defined transnational advocacy networks above, there are still some conceptual components of the definition that merit clarification.

The standard sociological concept of network refers to relations established among individuals to influence and constrain behavior on a certain issue or set of issues.[16] Usually, network members exhibit intellectual and emotional commitment to the issues at stake and share knowledge about them.[17] While expert knowledge and emotional commitment might explain why certain actors participate in an activist network,[18] I argue that they are insufficient to explain network participation by all types of actors. When it comes to understanding the participation of grassroots groups in transnational environmental advocacy networks, for instance, the notion of material interests has to be brought into the explanation. That is not to say that rural workers' associations, peasant cooperatives, and indigenous groups do not operate on principle or do not hold important knowledge on environmental issues. The point is that since they tend to be directly affected by changes in the local environment, they have a material interest in preserving their way of life and/or pursuing the betterment of their quality of life through environmental preservation.[19]

In the particular case of transnational environmental advocacy networks, I suggest that both ideal and material interests concur to explain the behavior of network members. They also help clarify conceptual differences among them. Transnational environmental networks are composed, primarily, of nongovernmental organizations.[20] There are, however, myriad definitions of NGOs. For some, they are "self-governing, private, not for profit organizations that are geared toward improving the quality of life of disadvantaged people."[21] For others, NGOs and interest groups are interchangeable terms, both defined as "private (i.e., nongovernmental) bodies organized for the purpose of directly or indirectly influencing public policy either on behalf of their members or on behalf of what they perceive to be the broader public interest."[22] Although these definitions are not contradictory, they emphasize very different aspects of what constitutes an NGO. In the first definition, NGOs are about improving the quality of life of sectors of the population, and we may assume that churches, and social assistance and self-help groups exhaust the categories in the concept. In the second definition the emphasis is on the political role of NGOs, and the fact that the term is equated to interest groups implies the inclusion of a broader range of private advocacy organizations (maybe even business associations and lobby groups).

The "overinclusiveness" of NGO definitions is detrimental to an accurate understanding of the composition and nature of activism in transnational environmental advocacy networks. In this book, I conceive NGOs (local, domestic, or international) as a different set of actors from grassroots groups.

NGOs are thus research or advocacy organizations that may provide support to grassroots groups at material and strategic levels but are not identified by the rank and file of such groups as co-participants in their political and material struggles. Several characteristics separate NGOs from grassroots groups: NGOs are usually professionally organized and have headquarters, communication resources, and permanent staff. They have specific mandates defined in statutes and cannot easily depart from them if, for instance, the objectives of a campaign in which they are involved suddenly change. Principles and values usually determine the priorities in their statutes. Despite the support they may provide to grassroots groups, NGOs rarely have a mandate to represent such groups. Examples of NGOs that participate in the networks discussed in this book are the Washington, D.C.-based Environmental Defense (EDF), and the Brazilian Institute for Amazonian and Environmental Studies (IEA). In contrast, grassroots groups may or may not have formal headquarters and paid staff. They are often (but not always) informally organized, and their membership tends to be restricted to those directly affected by the issue that originated concern and mobilization. Grassroots representatives have a formal mandate to represent a given population or social group.[23] While some grassroots groups may have statutes, these do not predefine issues or priorities for activism. Activism is determined by the needs of constituencies, which often change over time. Examples of grassroots groups are the various Amerindian regional associations and national confederations in Brazil and Ecuador, and the Organization of Rondonian Rubber Tappers. An important commonality between NGOs and grassroots groups is the one highlighted in McCormick's definition: both are political actors who directly or indirectly attempt to influence policy and politics at local, national, and international levels.

If the members of transnational environmental advocacy networks are of different natures, it is fair to assume that they have different relative impacts on a network's performance. I concede that international, and to a lesser extent, domestic NGOs tend to be the primary sources of material and technical resources for a network, and often take the crucial responsibility of producing and disseminating information.[24] Yet it is a network's local membership base that bears the responsibility of guaranteeing (through either direct implementation or monitoring) the eventual success of the network regarding the protection of the local environment.

As I give particular attention to the role of local groups in transnational environmental advocacy networks it is impossible to avoid an analysis of what such groups represent to the local civil society. Distinctions between local, national, and global civil societies, however, were not common in the literature until recently. Assessments of the degree of civil society activism were traditionally limited to nations. Alexis de Tocqueville, for instance, praised the role of peoples' associations and volunteer groups as constituting the backbone of

American democracy. Others highlighted the role of authoritarian political regimes in hindering the emergence and/or consolidation of national civil societies in Latin America and other regions of the world.[25] Minimally, civil society is always defined in contrast to the state. But beyond the boundaries of the family and clan and short of the state there is a good deal to be found: markets, voluntary associations, churches, interest groups, labor unions, non-governmental organizations.[26]

This array of actors broadens even further when one releases the concept of civil society from its national confines. Such a process became inevitable in the last decades of the twentieth century due to the political, social, and economic trends that reshaped the world during that period. The expansion of free-market economy, the information technology revolution, and processes of political opening and democratization of authoritarian and totalitarian regimes, have had direct effects in stimulating civil society organization at the local level, strengthening it at the national level,[27] and consolidating it at the global level. The very existence of transnational advocacy networks corroborates the notion that the phenomenon of a global civil society is real and is here to stay.[28] In fact, the globalization of information processes and technologies has been a crucial factor in the reorganization of power relationships at all levels of politics.[29] Social groups, traditionally marginalized by conventional (national) politics, have relied on information technology to project their plight and struggles beyond national borders. Thus, they have not only acquired allies and resources at the global level, but also transformed local demands into transnational ones. Identity-based movements, such as those of rubber tappers[30] and indigenous peoples,[31] have been particularly successful in using symbolic appeals and information campaigns as links between local and global activism.

In considering civil society at local, national, and international levels it is important to be mindful that "the concept of civil society does not make a smooth transition from the domestic to the international sphere if one expects them to have identical characteristics."[32] Thus, I must clarify what characteristics of civil society apply, equally, to all three levels. The first such characteristic is the diversity of groups and interests. The importance of transnational advocacy networks as a methodological tool is that they permit the identification of civil society groups that, despite their differences, obtain a certain degree of unity in pursuing a "common good." Other characteristics of civil society that transition well between levels of analysis include its being a space for the development of a community value system, and the fact that its functioning depends on association, communication, and information flows.[33] Finally, in an apparent but not actual contradiction to the two previous characteristics, civil society is an arena for conflict. At the local and national levels civil society is at odds with the state, attempting to assert its autonomy or complete separation from it. At the global level, civil society confronts the interstate system and the global economy. In any case, the tensions between public and private realms do not prevent their interpenetration. In the end, the boundaries between state (or the interstate system) and civil society are elusive and porous and actions in one realm have consequences for the other.[34]

While I do not refer often to the "global civil society" in the book, it is important to clarify that my approach to transnational advocacy networks assumes the existence of a "slice of associational life that exists above the individual and below the state, but also across national boundaries."[35] I do refer often to "local civil society" and more specifically to "local civil society groups." The difference between these terms is particularly important for an accurate assessment of processes of empowerment. One can, albeit with difficulty, "measure" the level of empowerment of certain groups in society over time. It is much harder, however, to evaluate processes of empowerment of whole civil societies (local or otherwise).

Applied to the specific cases of this book, local political and technical empowerment directly correlates to the political and technical capacity of local members of a transnational environmental advocacy network to promote environmentally sustainable development. Specific indicators of this process derive directly from the definition of empowerment provided above. Thus, one must assess the extent to which local groups have achieved a position of legitimate interlocutors vis-à-vis the state and other political and economic elites who have privileged access to local environmental policymaking processes; the extent to which local groups have guaranteed their access to policymaking arenas through formal channels that do not depend on specific activists or enlightened politicians and have used such channels to effectively influence the design and implementation of policies; the extent to which local groups have access to information on public policy and capacity to disseminate it among their rank and file; and finally, the extent to which participation in transnational activism has contributed to an increase in local groups' material and technical resources at adequate levels to meet the demands of participating in the formulation, implementation, and monitoring of environmentally sustainable development policies.

At this point, the reader may legitimately ask: "Why, then, does one need to discuss the promotion of environmental sustainable development by transnational environmental advocacy networks rather than placing the investigation squarely within the realm of local participatory development?" Recent critical assessments of both the practice and the theory on development and environmental resources management provide the answer. Local empowerment and civil society organizations' capacities to affect environmental policy occur neither in a political vacuum nor in an isolated socioeconomic context. The tendency to romanticize the "local" has skewed analyses away from acknowledging the inequalities and power relations inherent at that level, and from the broader national and transnational political and economic forces that affect local power imbalances.[36] Once again, the concept of transnational advocacy networks as a methodological tool, particularly when networks are investigated from the "inside out," sheds light on the interplay of power relations at various levels of analysis and on how these relations affect efforts to promote environmentally sustainable development.

Failure to elaborate on the tensions and cleavages that emerge among civil society groups, both locally and transnationally, may hinder the methodological relevance of transnational advocacy networks.[37] This is particularly true when one recognizes the need for activists to "negotiate over the terms of the story," or the "meaning" of their struggles and goals.[38] In the case of transnational environmental advocacy networks, the challenge of defining their struggles and goals is all the more complex due to the fuzziness of the concept of environmentally sustainable development.[39]

While the main "issue" binding together actors in the transnational advocacy networks discussed in this book is the promotion of environmentally sustainable development, not all network members approach this notion in identical ways. This should not be a surprise given that the term has been the object of debate in both academic and professional arenas, particularly since 1987, with the publication of the report *Our Common Future* by the World Commission on Environment and Development. The emphasis of the report was on the preservation of natural resources for future generations.[40] Such a broad definition had the somewhat positive effect of creating consensus among a wide array of actors. It played a role, for instance, in fostering a certain degree of unity among world leaders in the 1992 United Nations Conference on Environment and Development (UNCED), and in the formulation of guidelines for global action in areas such as biodiversity, water resources, and climate change. The major problem with that notion of sustainable development, however, was—and still is—its "fuzziness" or vagueness, and as a result, the difficulty in making it operational.[41]

A brief survey of the literature on environmentally sustainable development identifies at least three approaches to the concept. The first approach is less significant for the purposes of this book, since the environment is a secondary and implicit consideration in the larger context of "sustainable development." I will label this first approach "techno-economic." The other two approaches are labeled "mainstream" or "conservationist," and "socioenvironmental development."[42]

As its label indicates, techno-economic approaches to sustainable development rely on economic growth and technological advancements as key components of the process. In other words, "economic growth can create the capacity to alleviate poverty and solve environmental threats."[43] Such an understanding has followed Norgaard's (1984) proposal of linking economic and ecological paradigms, whereby sustainable development would be a possible outcome of a "co-evolutionary" (and preferably parallel) improvement of both economic and environmental systems.[44] Not all definitions of sustainable development within the techno-economic approach rely exclusively on technology and economic growth for the achievement of environmental sustainability. Admittedly, they are the key tools for the implementation of sustainable development in poor societies, but have limits when it comes to ensure equity within and among generations.[45] One alternative is thus to introduce the notion of "long-term" in the economic analysis. The concept of natural capital stock[46] may be one tool in this process. It would help in attenuating the dichotomy between development and environmental preservation. Traditional economic principles prescribe that environmental degradation actually increases the economic value of the next unit of environment since scarcity raises prices. Poor countries in particular tend to compromise their future development and the well-being of future generations due to immediate pressures to speed up development at the cost of compromising their environment. If a "long-term" perspective becomes predominant in economic calculations, the value of conserving a nation's natural capital stock could increase.[47]

"Mainstream"[48] or "conservationist"[49] approaches to environmentally sustainable development are placed together here only for the sake of brevity. In fact, they represent the largest and most diverse group of definitions of the concept. What mainstream and conservationist approaches to (environmentally) sustainable development have in common is their rejection of technology and traditional economic growth as the primary solutions for problems of environmental degradation. Not all definitions within this approach, however, reject the emphasis on technology and economic growth with the same intensity. Most admit that the elimination of poverty (through these processes) is an essential condition for environmental protection. What characterizes mainstream and conservationist approaches to environmentally sustainable development is not only that, compared to techno-economic approaches, they underplay the role of technology and economic growth in the process, but also that they rely on other variables. In this sense, environmentally sustainable development is a process in which not only economic growth matters, but one in which the quality of growth is paramount. Quality of growth is dependent upon the control of population levels, the conservation and enhancement of the natural resources base, and the participation of all stakeholders in decisions regarding environmental preservation and sustainable development.[50]

Finally, socioenvironmental development approaches emphasize the ideals of equity, social justice, and political participation as inherent components of environmentally sustainable development.[51] One of the main assumptions of socioenvironmental development approaches is that "the way people relate to their environment—as well as the way they understand it—is created by culture, and bounded by social relations, by structures of power and domination."[52] Hence, sustainability requires looking beyond the natural environment per se, and toward a political economy approach to environmental problems.[53] This is a significant departure from mainstream or conservationist approaches, which tend not to challenge existing social, economic, and political structures, but suggest reforms in the ways these structures affect the natural environment. Analyses of the environmental crisis from a socioenvironmental (or political economy) perspective are mindful of the need for poverty alleviation if environmental sustainability is to be achieved.[54] Yet, different from mainstream and techno-economic approaches to environmentally sustainable development, these analyses totally reject the notion that economic growth will eliminate poverty. On the contrary, economic growth is more likely to be a cause of increasing levels of socioeconomic and political inequalities.[55] Without a radical change of structures and processes that perpetuate socioeconomic and political inequalities, environmentally sustainable development cannot be achieved. It is interesting to notice that a socioenvironmental development approach to environmentally sustainable development seems to have left the more radical periphery of environmental analyses to become predominant among renowned students of environmental and development issues in Amazonia.[56]

As the stories in this book unfold, the reader will have the opportunity to observe how the different approaches to environmentally sustainable development influenced the actions of transnational environmental advocacy networks and of the different actors who participated in them.

Key Questions

The main part of this book consists of a comparative study across time of a particular environmental advocacy network, the Rondônia network. The Rondônia network emerged in the early 1980s and mobilized environmental and human rights international NGOs, environmental activists and consultants for environmental and Amerindian issues both in Brazil and abroad, the specialized media, and concerned individuals in multilateral and governmental agencies. These individuals and organizations had in common their concern with the environmental consequences of development policies then under implementation in the Brazilian state of Rondônia, in western Amazonia. The analysis of the evolution of the Rondônia network over a period of twenty years (1980–2000) illuminates, in an unprecedented way, the challenges and opportunities confronting transnational environmental advocacy networks.

Theoretical and practical motives determined the selection of the Rondônia network over other possible choices. For reasons that shall be detailed in the following chapters, the Rondônia network generated, from its onset, high levels of interest among the global environmental and human rights communities, and at specific moments, among the general public as well. As a consequence, it has become a landmark of transnational environmental activism. In addition, its time span (twenty years) allows for conclusions that address structural, rather than circumstantial issues. Finally, research on the Rondônia network was made easier by my personal and professional contacts in Brazil and fluency in Portuguese. For comparative purposes, I also studied, in significantly lower levels of detail, transnational advocacy networks in Ecuador and India (see chapters 6 and 7). The selection of these networks followed the theoretical rationale presented above, namely, the fact that they generated significant levels of interest worldwide and eventually became landmarks for transnational social and environmental activism, and their long time span (beyond the scope of specific campaigns). To focus the analysis I resorted to several questions about the nature of transnational environmental advocacy networks and the impacts of their activism.

Who participates in a transnational advocacy network and how do they participate?

At first glance, this is more an empirical than an analytical question. In truth, it is not. As one investigates the composition of a network he/she inevitably evaluates the relative weight of network members. Different political and material resources, differential access to political arenas, different sources of legitimacy, and different roles in decision-making processes affect relations among members of a network. These internal relations are determinant of a network's effectiveness. As this book unfolds, the reader will notice that local groups' membership in transnational environmental advocacy networks does not automatically guarantee their meaningful *participation* in them. Unless local groups devise or create avenues through which their priorities and "vision" are incorporated into a network's overarching goals, they risk becoming mere instruments of legitimation for international environmental activism.

At the onset of each chapter I list the organizations and groups of activists that were most active in the networks (noting when and if the relative weights of different players within each network change overtime). In doing so, I describe network members' characteristics, resources, and goals. As the chapters unfold, the reader will find answers to questions such as: How did network members negotiate the terms of their common struggle (the meaning and goals of their mobilization)? How did network members reach decisions about specific strategies? And what role did each member or group of members play in this process? The internal political exchanges among network members may provide clues for an improved understanding of multilevel politics beyond that of specific transnational networks.

What are the strategies available to transnational advocacy networks? When are they successful and why?

The study of strategies devised and implemented by a network as a whole and/or by some of its members at specific junctures constitutes an important part of the explanation for a network's successes and failures. This aspect of the analysis is of particular interest for practitioners and activists. Evidently, a given strategy, used to pursue a specific goal, in a given moment in time, is a historical experience that cannot be replicated. Yet, understanding the conditions in which a given strategy was more or less successful may provide valuable insights for ongoing and future struggles.

In evaluating the strategies used by the members of the networks discussed in this book, I looked for answers to the following questions: What were the objectives of specific strategies and how did they relate to both the overall goals of the network and to the specific agendas of particular members? Who were the key catalysts for such strategies within the network? Who were the primary targets of specific strategies and how did such targets react? The case studies will reveal the effectiveness of locally devised and locally implemented network strategies, despite the tendency of network members to privilege initiatives that unfolded in the international arena.

What are the consequences of transnational advocacy networks' activism?

This question goes to the core of the theoretical ambition of this book. The consequences of transnational advocacy networks must be evaluated in three different areas. First, there are the consequences for network members themselves. How did their experiences and involvement in a given network affect their material resource base, political alliances, legitimacy vis-à-vis their constituencies, and assessment or reevaluation of goals? How did the evolution of a network over time affect the balance of political forces among its members, and conversely, how did changes in this balance of forces impact a network's effectiveness? Many of the answers to these questions turned out to be counterintuitive. They provided a foundation for my challenge to the assumption that the mere participation of local groups in a transnational advocacy network leads to their political and technical empowerment. In fact, the effort of joining transnational activism may, on occasion, lead local activists and local organizations to overstretch themselves, attempt to shoulder burdens beyond their technical and political capacities, and acquire a level of exposure that may prove detrimental to the long-term sustainability of their struggles.

The second area of consideration about consequences of transnational advocacy networks must address a network's specific goals. Were they accomplished as a result of network activism? Did they have to be redefined as a result of unforeseen obstacles (or opportunities)? At what costs? In the specific case of transnational environmental advocacy networks, what are the consequences of their successes or failures for theoretical and practical approaches to environmentally sustainable development? Here I hope to advance the notion that the concept of environmentally sustainable development is all the more useful as it is approached as context-dependent, rather than as a vehicle for uniformity and consensus.

Finally, what are the consequences of transnational advocacy networks for the political contexts in which they operate? The focus of this book is on the impact of networks on local politics, particularly to the extent that they affect the level of political and technical empowerment of local civil society groups. Yet the book also discusses the consequences of transnational activism at the national and international levels, such as changes in national and international policies and the creation and reformulation of international mechanisms for grievances (such as the World Bank–sponsored Inspection Panel).

Research Method and Organization of the Book

I used several research methods to conduct this study of transnational advocacy networks. The history of the networks was reconstituted both from secondary sources and open-ended interviews. For the evaluation of the environmental challenges that the networks confronted and of the specific environmental impacts of networks' strategies I relied on technical sources such as reports by independent consultants, environmental NGOs, national governmental agencies in charge of policy implementation, and the World Bank.

Networks' politics and impacts on members and on the local political context were inferred from the analysis of documents from the archives and websites of network member organizations (such as correspondence among activists, summaries of mobilization strategies, reports of field trips, and memoranda of meetings), articles in local and international newspapers, and open-ended interviews.

I conducted more than sixty interviews during a ten-year period (1991–2001) in Brazil, Ecuador, and Washington, D.C., with NGOs and grassroots activists, government representatives at local and national levels, World Bank staff, consultants for environmental and Amerindian issues, and officials in private sector associations. I did not attempt to obtain a numerical balance among the interviewees based on their institutional affiliation (government official, NGO/grassroots group representative, or staff at a multilateral organization) or level of activism (transnational, national, or local). I looked for individuals that were most directly linked to—or affected by—the transnational advocacy networks focused upon in this study. I must confess that, except for the logistics of traveling long distances and for extended periods of time, I did not encounter any significant difficulty in conducting interviews. Initial fieldwork coincided with the preparation and immediate aftermath of the 1992 United Nations Conference on Environment and Development. I believe that this research benefited, in part, from the interest generated by that event. Whenever I presented the theme of this study, I was greeted with a positive response from the potential interviewee based on his/her own interest in the topic and acknowledgement of its importance. I was also favored by the fact that close to two-thirds of the interviews were conducted with the primary goal of obtaining data for my dissertation (of which this book is a by-product). My condition as a Ph.D. student engaged the sympathy (and sometimes the pity) of interviewees who had once experienced the trials of graduate studies (many World Bank officials, consultants and staff in research institutes and international NGOs). Being Brazilian most certainly contributed to the level of comfort of my conversations with Brazilian activists and government representatives and with leaders of Rondonian civil society groups (the point was explicitly made by more than one interviewee). Finally, most interviews were conducted in Portuguese and in English, languages in which I am fluent. The interviews in Spanish were conducted with the help of a research assistant fluent in that language.

Before initiating the analysis of transnational environmental advocacy networks in Brazil and beyond I provide, in chapter 2, a historical background on development and environmental protection initiatives in the Brazilian Amazon region. The chapter describes national and international policies devised for the region from the mid-1960s to date. Development in the state of Rondônia and the environmental consequences of this process is discussed in relation to this larger context. The chapter highlights how economic, financial, and political demands of the national and international contexts impacted on the local and regional environments.

In chapter 3 I describe the origins of the Rondônia network in the early 1980s to mitigate the environmental and social impacts of highway construction and colonization in the state. I explore the dissonance of goals and choice of strategies among international and national members of the network, and discuss how these problems affected the network's impact and evolution.

Chapters 4 and 5 describe the evolution of the Rondônia network in the 1990s. In chapter 4 I analyze the network's effectiveness in influencing the design of the Planafloro project, an internationally financed program to manage Rondônia's natural resources. I discuss how the network's effectiveness was affected by efforts to deepen its local membership base and the consequences of this process for both the network and the local environment. Finally, in chapter 5, I evaluate the Rondônia network as it reached its political maturity. The chapter describes efforts to overcome legitimacy challenges that affected the network in the early 1990s, and evaluates the impact of specific strategies in this process. The main focus of the chapter is on the role played by local groups in the Rondônia network at the turn of the millennium and the consequences of local activism for local politics and the environment.

Chapters 6 and 7 offer an opportunity for comparison between the trajectory of the Rondônia network and those of Ecuador's anti-oil network and India's Narmada network (emphasizing this latter's campaign against the Sardar Sarovar hydroelectric project). The focus of these chapters is on the effects of participation in transnational activism for the empowerment of local civil society groups and the protection of their natural environments.

2

The Dilemma of Amazonian Development and its Impact on Rondônia

This chapter summarizes the opportunities and contradictions of development and environmental policymaking in Amazonia from 1964 to the present. It focuses on the political and socioeconomic factors, both domestic and international, that have shaped policies and the allocation of state resources for the region. In doing so, it introduces the larger context in which the Rondonia transnational environmental advocacy network has operated from its inception, in the early 1980s, to date.

In Brazil, one often refers to gigantic realities by using the expression "of Amazonic proportions. . . ." In Amazonia everything amounts to huge numbers. The area defined as Legal Amazonia covers 58 percent of Brazil's national territory (more than 5.0 million square kilometers).[1] The world's largest river in volume of water—the Amazonas River—runs through it. Amazonia is the site of the world's largest area of continuous tropical forest,[2] which is estimated to host thirty thousand species of plant life, while more than three thousand species of fish inhabit its rivers.[3] The region is also believed to contain significant reserves of mineral resources, a belief supported by the discovery of the Carajás mineral deposits—the world's largest reserve of high quality iron ore.[4] Finally, the sheer size of the region breeds hope that some of its soil would be appropriate for agricultural production, a prospect that has become increasingly controversial.

Given such characteristics, it is not surprising that Amazonia has become a paradigmatic example of the dilemmas between development and environmental conservation. The history of development initiatives in Rondônia, an area in Amazonia originally defined as a federal territory, but elevated to statehood in 1982, is particularly illustrative of such dilemmas. Rondônia's pattern of development has been constrained by larger development policies for Amazonia, which I describe below.

The year 1964 marks the inauguration of a military authoritarian regime in Brazil that lasted until 1984. Both domestic and international pressures influenced the military's plans for the colonization of Amazonia, perceived as Brazil's "last frontier." Domestically, geopolitical and national security objectives required the presence of Brazilians and the Brazilian military (as opposed to that of indigenous peoples or foreign missionaries, for instance) in remote frontier areas. Border control was thus to be achieved through the promotion of settlement schemes, the establishment of a military presence, and the construction of roads in the jungle. The occupation of frontier areas was a major tenet of the ideology of "national security," developed by Brazil's highest institute of military education.[5] The Brazilian military in the 1960s worried about possible expansionist intentions of neighboring nations and infiltration of Cuba-sponsored communist insurgents. Rondônia, located in the northwest region of Brazil, raised special concerns. Its western limits coincide with Brazil's frontier with Bolivia, a country that in the 1960s was immersed in political unrest and whose leftist groups were receiving direct assistance from Castro's Cuba.[6]

International pressures for the colonization of Amazonia were economic in nature. By the mid-1970s, the military's economic project for Brazil was faltering due to several factors: the 1973 oil crisis and subsequent world recession, the increase in Brazil's foreign debt, and the unfavorable international prices for Brazilian exports. Brazil hoped to avoid a worsening of the crisis by increasing and diversifying its export base. Amazonian natural resources and agro-industrial potential were key assets in this strategy. The military's first attempt to address its geopolitical and economic concerns regarding Amazonia was a set of policies under the rubric *Operação Amazônia* (Operation Amazonia). It was followed by two other macro-policy schemes that emphasized the economic growth of the region while entirely disregarding related environmental impacts. I discuss the details of such policies in the first part of this chapter.

In the second part of the chapter I describe environmental and development policies for Amazonia that were devised after Brazil's return to democratic normality in the late 1980s. Increased political freedom and the intensification of international concerns over the fate of the Amazon forest marked that period. These factors contribute to explain a change in Brazil's official rhetoric about Amazonia and the inclusion of the phrase *environmental sustainability* in the official discourse of both the federal government and the administrations of Amazonian states. Ironically, changes in the official rhetoric coincided with one of Brazil's worst economic crises, which lasted from the late 1980s to the early 1990s, and placed further pressures on Amazonian natural resources. Public policy for the region was marked by contradictions.

For the sake of clarity and brevity, this chapter summarizes only the most significant federal policies for Amazonia between 1964 and 2001. It does not discuss specific environmental and development policies of individual Amazonian states. State policies have been constrained by federal initiatives, as will become evident as I examine the specific case of Rondônia.

AMAZONIAN DEVELOPMENT AND MILITARY RULE (1964–1984)

Operação Amazônia, the first set of policies for Amazonia devised by the military, had two major objectives: the integration of the region into the national economy, and the physical occupation of the area. Specific institutions were created to foster these processes, namely, a regional development agency and a regional bank.[7] These institutions primarily aimed at attracting private enterprise to Amazonia, mainly through the concession of tax incentives and special credit lines. In addition, *Operação Amazônia* targeted the region's infrastructure. The Belém-Brasília highway, connecting one of the most important Amazon cities (Belém) to Brazil's capital (Brasília), was completed in the mid-1960s. The highway opened the region to immigration by small farmers as well as to large logging and ranching enterprises.[8]

To further encourage Amazonia's occupation, the Brazilian government extended fiscal incentives and credit to economic initiatives in the area. Before 1964, legislation only allowed individuals to deduct up to 50 percent of their income tax for investment in agricultural and industrial initiatives in Amazonia. From 1966 on, tax exemption was extended to corporations, and applied to investments in any sector—industry, agriculture, livestock, or services.[9] Amazonian states matched the incentives from the federal government, increasing the resources available. Finally, as an indirect strategy to expand incentives, federal legislation mandated that land acquisition be considered as a part of development costs. As a consequence, land transfers became exempt from taxation.[10] *Operação Amazônia* received the support of the international finance community in the form of loans by multilateral banks, such as the World Bank and the Inter-American Development Bank.

The Brazilian system of tax and credit incentives, as well as regulations on land allocation and titling, significantly affected Amazonia's development in the 1960s and 1970s. In an outstanding study, one researcher dissected Brazilian legislation and revealed its impact on Amazonia's human and natural environments.[11] He explains that, during the military rule, between eighty and ninety percent of profits from agricultural and livestock initiatives were exempt from taxation. Investment in agriculture or livestock was thus inherently profitable, regardless of its actual rate of return.[12] In Amazonia, in particular, specific legislation provided income tax holidays, tax credits to corporations establishing business in the area, and in some cases, exemption from import tariffs, export taxes, and commodities taxes for imports and exports. These benefits, however, only applied to individuals and corporations that could prove legal ownership of the land. Both land rights and titling procedures in frontier areas, such as Amazonia, however, are extremely complex issues, as many studies have shown.[13] This complexity favored those who could count on legal advice and knowledge of the system, that is, corporations and wealthy individuals, while punishing the less resourceful, such as small farmers and landless migrants. In addition, the system has contributed to environmental degradation. Until recently, a claimant could expedite land titling procedures if he/she presented evidence of productive use of the land. Land agencies consistently accepted land clearing and the conversion of forest into pasture as indicators of land productivity.

In an attempt to reverse some of the distortions in the patterns of land distribution and occupation that *Operação Amazônia* accentuated, as well as in response to increased population pressures in the northwest of Brazil, the military conceived a new policy. The National Integration Plan (PIN, 1970–1974), while maintaining fiscal incentives and subsidized credit for agribusiness and livestock operations, created new policy instruments that aimed at a systematic integration of small farmers into the colonization effort. The new policy also aimed at expanding the range of economic activities in Amazonia, and there were funds earmarked for the expansion of infrastructure for mineral exploitation.[14]

Road construction was a key strategy in promoting PIN's objectives. The two most ambitious projects designed and partially implemented during this period were the Trans-Amazonia Highway, a five thousand-kilometer road linking the northeast of Brazil to the state of Amazonas, and the paving of the Cuiabá–Porto Velho Highway (BR-364). The latter became the cornerstone of Rondônia's development and of the environmental degradation that ensued. Road construction, marketing campaigns, and package incentives to potential colonists from Brazil's south and northeast regions led to uncontrolled immigration to Amazonia. In the early 1970s, migrants settled mainly along the Trans-Amazonia and Belém-Brasília highways, while large agribusiness projects were directed to the area of the Cuiabá–Porto Velho Highway. By the end of the decade, however, Rondônia had become a primary destination for migrants, who originated not only from outside Amazonia, but also from within it. The government's inability to control migration and promote effective settlement programs led to PIN's failure. In part, such an outcome was the result of policymakers' environmental shortsightedness. As one researcher explains, the routing of highways ignored crucial issues such as soil fertility and topography, the economic and health-related consequences of disrupting the environmental balance of the region, and the fact that manmade deforestation would inevitably become the last resource for poor farmers to increase soil fertility.[15]

Both internal and external constraints led the Brazilian government to reevaluate Amazonia's development policies. The 1973 oil shock and the world recession that ensued forced Brazil to further rely on its export base as a means toward the stabilization of the commercial balance. While in the 1970s, the government prioritized grain and beef exports, in the early 1980s it focused on minerals, which originated primarily in the eastern Amazon region of Carajás. Besides the constraints of the world economy, Brazilian policymakers were under domestic pressure from entrepreneurs—mainly the Amazonian Association of Agriculture and Ranching—who resented PIN's priority on small farmers. It did not take long for PIN to be formally abandoned, and in its place the Brazilian government promoted the development of "growth poles" (or centers of growth) throughout Amazonia. This strategy became known as Polamazonia (1974–1983).

Polamazonia deepened previously existing incentives for private capital—particularly agribusiness and mining—to invest in the region. The link between Brazilian tax and credit incentives, land ownership structure, and environmental degradation that research has unveiled,[16] emerges clearly in the context of this new scheme. In fact, the Amazonian regional development agency, SUDAM, after 1975, limited eligibility for fiscal incentives to ranching projects of more than twenty-five thousand hectares. Concurrently, the growing inflationary trends of the Brazilian economy made land an important resource against capital depreciation. Finally, the lack of credit incentives to small farmers led to a high turnover in land ownership. While wealthy landowners acquired land from indebted farmers at bargain prices, thus furthering land concentration, landless migrants moved deeper into unclaimed areas of the forest (namely, public forests and Amerindian land). Although direct causality cannot be established, data from several sources suggest a link between the processes of increased credit incentives, land concentration, and deforestation. For instance, the number of agribusiness projects receiving government incentives in Amazonia rose from 144 in 1970, to 209 in 1980, and 359 in 1983.[17] During the same period, land concentration, measured by the number of properties of more than twenty-five thousand hectares in five Amazonian states, rose between one and twenty percentage points.[18] Finally, while in 1975, the percentage of area cleared in Amazonia varied between zero and two points, in 1980, many states were close to the five percent mark. In 1988, three states had broken the fifteen percent mark, with Rondônia being close to twenty-five percent.[19]

In the context of Polamazonia, the colonization of the Amazonian northwestern frontier, and particularly of the region known as Rondônia, gained momentum. The Polonoroeste project (1981–1987) became the cornerstone of this process. Polonoroeste is the Brazilian acronym for the Northwest Brazil Integrated Development Program that Brazil presented to the World Bank for evaluation in 1979. The core task of the program was the paving of the Cuiabá–Porto Velho Highway (BR-364), built in the 1960s as a dirt road. There were also other projects within the Polonoroeste program—the establishment of new colonization schemes in Rondônia and the consolidation of existing ones, the provision of health infrastructure, the establishment and enhanced control of national forests and ecological stations as well as other measures for environmental protection, and the protection of local Amerindian populations.[20]

Brazil used three main arguments to justify the Polonoroeste project to Bank officials: 1) the project would improve the area's transportation system, an essential element for the promotion of colonization initiatives; 2) the project would develop the agricultural potential of the region, which in contrast to most of Amazonia, was believed to have climatic and soil conditions suitable for agriculture; and 3) the project would help control the unusually high rates of migration to Rondônia, which had been rising since the opening of the Cuiabá–Porto Velho dirt road, and were already presenting serious environmental and social risks.[21]

During 1980 and 1981, the World Bank carried out a series of analyses of Polonoroeste in order to evaluate its risks and potentials. It concluded that although the project could transform Rondônia into an important agricultural and timber-producing region in Brazil, its execution would entail a higher than normal degree of risk. Risks derived from the region's rapidly growing population, the consequently confused land tenure situation, the area's fragile and imperfectly known natural environment, the condition of local Amerindian populations in early stages of contact with modern society, and, finally, the thin administrative structure in place.[22] World Bank analyses warned that the Brazilian government should be prepared for the negative consequences of accelerated development in a frontier area, such as violent conflicts over land, invasion of Amerindian reserves, deforestation, and lawlessness. Yet the World Bank believed that a strong central control and the inclusion of special measures in the Polonoroeste program could attenuate these impacts.

On December 1, 1981, the World Bank and the Brazilian Government signed the Polonoroeste agreement, whereby the Bank would finance about one third (US$443 million) of the total cost of the project. The agreement, however, was a significantly modified version of Brazil's original proposal. The World Bank conditioned its participation in the Polonoroeste on a series of measures. Among these measures, several related to environmental and Amerindian protection and were summarized as: 1) allocation of funds for the creation of biological reserves, a national park, four ecological stations, and several national forest areas; 2) demarcation of fifteen Amerindian reserves as well as the protection of Amerindians' health; and 3) concession of land to small farmers on fertile soils, thus preventing invasion of Amerindian land.[23]

Two years after the formal initiation of Polonoroeste, it was already evident that neither the Brazilian government nor the bank could keep control over the project's impacts on Rondônia's society and the environment. Although bank staff had predicted that the thin administrative structure in the Brazilian northwest would be a source of problems, its consequences were underestimated. Difficulties derived not only from the absence of institutional mechanisms to implement most of the project's provisions, but also from the weakness, malfunctioning, and total lack of effectiveness of existing institutions. The most often cited examples of this institutional failure were the Brazilian Indian agency—the *Fundação Nacional do Indio* (FUNAI)[24]—and the environmental agency—the *Instituto Brasileiro de Desenvolvimento Florestal* (IBDF). Structural problems nevertheless existed in almost every federal and state agency involved in the program's implementation.

Among international audiences concerned with environmental issues, the Polonoroeste soon became the best-known example of an unsustainable development scheme. As I discuss in the following chapters, the project was the target of criticism from several sectors of the Brazilian and international civil societies. Environmental and human rights groups, activists, and the specialized media initiated an international campaign demanding that both the World Bank and the Brazilian government be accountable for the negative impacts of Polonoroeste. In 1985, in response to such pressures, the World Bank suspended disbursements for the project. The initiative generated serious diplomatic tensions between Brazil and the bank. It also established a precedent, since it was the first time that a development project had been interrupted based on its bad environmental—not economic—record. The bank quickly reinstated disbursements in return for Brazil's commitment to tightening supervision of environmental and Amerindian-related aspects of the Polonoroeste project. Yet most of these initiatives were abandoned as soon as Brazil declared the project officially completed in 1987.[25]

The completion of the Polonoroeste project coincides in time with the final period of Brazil's transition from military to civilian rule. The process of democratization involved, among other things, the abandonment of the 1967 military-inspired constitution and the formulation of a new one. The promulgation of Brazil's 1988 Federal Constitution greatly reduced the level of centralization of development policies in Brazil. In the years that followed, states and municipalities acquired a higher degree of autonomy from the federal government in defining and drafting regional policies. The Amazon region has since become a stage in which development, environmental, and even security policies unfold, often in contradiction to each other. This situation is further complicated by the intensification of global trends that directly affect the area, and accentuate tensions between economic growth and environmental protection. Among these trends are the continuing international concern with the fate of the Amazon forest, the increased demands for natural resources by the global economy, and new security threats to the Brazilian northern and northwestern frontiers (namely drug traffic).

Environmental and Development Policies for Amazonia in Democratic Brazil

In mapping the policies for the protection and sustainable development of Amazonia in the last fifteen years one must consider four vectors. The first consists of environmental policies that are national in scope, but that have particularly significant impacts on the Amazon region. The second relates to internationally driven funding initiatives for environmental protection. The third is the proliferation of natural resources management projects in Amazonian states, illustrated here by initiatives in Rondônia. Finally, and to a certain extent in contrast with environmental policies, there are the continuing efforts to promote the economic development and security of the region. After a decade of being officially dormant, these efforts have regained momentum in recent years.

The information campaign in the international media about Polonoroeste's impacts, and the lobbying of environmental groups against the project in the U.S. Congress and European parliaments, greatly contributed to focus international attention on Amazonia. The scope of the 1987 fires, which surpassed those of previous years, further increased concerns. Finally, the assassination, in 1988, of Chico Mendes, the leader of Amazonian rubber tappers and an award-winning environmental activist, shocked the world and put a human face—that of forest peoples—on the efforts to save the Amazonian environment.

The Brazilian government deployed serious efforts, both in the domestic and international arenas, to discredit concerns with the fate of Amazonia. It disqualified such concerns as threats to the national interest and sovereignty. More conservative sectors even identified signs of a conspiracy to internationalize the region.[26] Eventually, however, the Sarney administration (1985–1989) launched an environmental plan for Brazil, with an emphasis on the Amazon region. The plan, strategically named "Nossa Natureza" (Our Nature), abolished subsidies and other incentives that encouraged deforestation, and suspended some development initiatives.[27] It also created the National Fund for the Environment. The fund (US$22 million obtained through a loan from the Inter-American Development Bank) was to finance environmental protection initiatives by municipalities and non-governmental organizations (NGOs). While *Nossa Natureza* marked the first attempt by the Brazilian government at formulating an encompassing environmental policy, the fund reflected the federal government's acknowledgment that environmental protection required decentralization and partnership with civil society organizations. Yet both *Nossa Natureza* and the fund quickly became mere diplomatic marketing tools rather than effective instruments to protect the Amazonian environment.[28]

In 1990, the newly elected president Collor de Mello became keenly aware of the importance of Amazonia's environmental protection for the international community. During Mello's international trip as president-elect, the topic emerged in conversations with most European chiefs of state. It was also the object of public rallies organized by environmental groups in many cities that the president-elect visited. Mello was quick to perceive the opportunities that such a level of interest could generate for Brazil. To launch a new image of Brazil as an environmentally committed nation, the Mello administration initiated intense lobbying efforts to host the 1992 United Nations Conference on Environment and Development. It combined these efforts with high-visibility actions, such as expelling gold miners from Amerindian reserves in Amazonia and bombing their illegal air strips.

The effort to create an environmentally friendly international image of Brazil contributed to the decision of international donors to finance two important environmental programs: the Pilot Program for the Conservation of Tropical Forests (Pilot Program), and the National Program for the Environment (PNMA). The Pilot Program originated from an offer by the G-7 Group to Brazil.[29] It consists of a large fund (approximately US$60 million), bilateral co-financing mechanisms, and Brazilian resources to be channeled to environmental projects. The World Bank is in charge of managing resources for the Pilot Program, which have been used to fund economic and ecological zoning efforts, monitoring schemes for protected areas, demarcation of forests, extractive reserves, and Amerindian lands, institutional strengthening, research, and environmental education. International donors expected that, if successful, the program could be replicated in other areas of tropical forests.[30] The success of the program, however, ten years after its formulation, is still open to evaluation. On the one hand, its resources have been instrumental in the demarcation of Amerindian and extractive reserves in Brazil and in Amazonia in particular.[31] Most recently, one of its components, the Demonstrative Projects, designed to integrate local populations and NGOs directly into the objectives of the program, has received praise.[32] On the other hand, the sheer size of the program creates problems of interagency coordination and bureaucratic delays in the distribution of resources.

The same problems have plagued the National Program for the Environment, but in an even more acute form. The PNMA was an effort to restructure the Brazilian environmental sector by strengthening IBAMA, the Brazilian institute for the environment. In addition, the program aimed at the creation and improved protection of conservation units and ecosystems. Its funding came from the World Bank in 1990, a time when the bank was eager to demonstrate its commitment to the environment, in part because of the criticism it received for its role in the Polonoroeste project. Brazilian fiscal and economic difficulties in the early 1990s, however, practically imploded the program.[33]

Despite the difficulties encountered with the PNMA, the World Bank continued to evaluate proposals for environmental funding in Brazil. In the 1990s, due to the Brazilian economic crisis, there was no shortage of such proposals. External financing was often the only alternative for cash-starved regional administrations to obtain a certain degree of liquidity. In addition, with the administrative decentralization created by the 1988 constitution, states had acquired a higher degree of autonomy to negotiate international loans directly with multilateral finance institutions. This was precisely what the state of Rondônia did.[34]

Even before the Polonoroeste project was officially completed, Rondonian and World Bank officials had started to discuss the possibility of developing a plan to address the area's environmental problems. It is debatable whether concerns with Polonoroeste's environmental devastation were the determinant factor behind such discussions. For the Rondonian state, the environment had never been a priority. Economic development, particularly in the form of incentives to extractive industries such as logging and mining, had traditionally been the focus of every state administration. The decision to push for an environmental project, in the absence of any structural change in the state's politics, seems to have been determined mainly by financial needs. On the part of the World Bank, new funding for Rondônia, in light of all the implementation difficulties of Polonoroeste, did not make much sense. Yet, in public relation terms, funding an environmental project in Rondônia was just the opportunity needed for the bank to further stress its new commitment to environmental issues.[35]

Riding on the momentum created by the UNCED 1992, the Word Bank and the Rondonian Government signed the Loan Agreement for the Rondônia Natural Resources Management project, or Planafloro, for its acronym in Portuguese. The World Bank provided most of the funding for the project (US$167 million, out of a total project cost of US$200 million). Planafloro's main goals were: 1) to promote a socioeconomic and environmental zoning plan in Rondônia, and subject to it all regional development initiatives;[36] 2) to promote institutional and public policies in Rondônia compatible with the zoning plan and the objectives of sustainable development; 3) to develop integrated farming systems in areas suitable for agriculture and agro-forestry; 4) to limit the expansion of agro-industry (mainly cattle ranching) in the state; and 5) to protect and enforce the borders of all conservation units and Amerindian reserves.[37]

Being one of a new generation of projects within the World Bank, Planafloro design emphasized decentralization. State agencies held the primary implementation responsibilities, although selected federal agencies were also involved, such as the National Indian Foundation, the Brazilian Institute for the Environment and Renewable Natural Resources (IBAMA), and the National Institute for Colonization and Agrarian Reform (INCRA). An interesting feature of Planafloro, and one that strikingly distinguished it from the Polonoroeste project, was the creation of a state council as the formal decision-making body for the project. The council was presided over by Rondônia's governor and included representatives of the federal and state implementing agencies, the association of the state's mayors, and civil society organizations.[38] As the following chapters will demonstrate, the participation of NGOs and grassroots groups in the Planafloro project is one of the most controversial, yet politically interesting, experiments of civil society's monitoring of environmentally sustainable development policies in Brazil, and possibly in the world.

While local civil society organizations were formally included in the Planafloro institutional structure, international environmental NGOs were the key players in influencing project design. Due to NGOs' pressures, the final version of the project contained a series of preconditions that included, among others, the completion of environmental and Amerindian protection measures that had been required—but remained incomplete—within Polonoroeste. Despite gains in the project design, the transnational environmental advocacy network that, since Polonoroeste, remained mobilized to protect Rondônia's environment, had little success at the onset of Planafloro's implementation. During the project's initial years, practically none of its environmental promises were fulfilled.

This scenario changed drastically in 1995 when the Rondonian advocacy network reached unprecedented levels of cohesion and devised innovative pressure strategies. Still, Planafloro remained far from being a successful example of an environmentally sound development program in Amazonia. Its promise was undermined by the resilience of the economic development model that, despite all attempts to the contrary, continues to shape Amazonia's destiny.

It should be no surprise, then, that, parallel to the highly visible and well-marketed environmental policies of the 1990s, the Brazilian government has continued to support traditional approaches to the region's development and security. While the Sarney administration bolstered its commitment to Amazonia and the Brazilian environment by formulating the *Nossa*

Natureza program, it sponsored *Calha Norte*, a project that further aggravated environmental and Amerindian problems.[39]

The *Calha Norte* project was a military initiative, drafted by the Secretariat of Strategic Affairs (SAE) with the collaboration of several ministries. Although the project did address environmental and indigenous issues in Amazonia, the environmental ministry did not participate in it. *Calha Norte*'s main goal was the protection of Amazonia, defined in broad terms. It included border defense and actions against internal and external sources of instability (increased numbers of land conflicts inside Brazil, and drug trafficking and guerrilla activities on the borders with Peru and Colombia), and the defense of natural resources (combating biotech piracy, halting timber and minerals smuggling, controlling forest fires and illegal mining).[40] It also defined border protection in terms of military presence, even if it implied displacing Amerindian groups or exposing them to unregulated contact with the evolving society (military personnel), and to all the risks that it entailed.

Its approach to Amerindians and their rights soon became the most controversial aspect of *Calha Norte*, both inside and outside Brazil. The Brazilian military have traditionally viewed the creation of Amerindian reserves in border areas as a geopolitical threat. According to them, Amerindians' autonomy increases the risk of secession. In addition, the constant and loosely controlled presence of foreigners (missionaries, activists, researchers) in Amerindian areas undermines national sovereignty. In the late 1980s, intense international pressures for the demarcation of the Yanomami land, the largest indigenous area in Brazil, which stretches well into Venezuelan territory, intensified such fears.[41]

Critics of the military's approach to Amerindians and of the *Calha Norte* project argue that the military were never interested in preserving the area's environment. The goal of the military has always been to limit access by small farmers and Amerindian populations to areas of mineral, livestock, and lumbering potential, thus setting them aside for exploration by parastatal companies.[42] Be that as it may, the project's goals of supervision and defense of Amazonia were never achieved. Such objectives required a level of resources that, in the midst of the economic crisis of the 1980s and early 1990s, the Brazilian government was not able to provide, even to the military.

Although *Calha Norte* remained dormant for most of the 1980s and 1990s, it has been recently resurrected in the wake of new initiatives for Amazonia. The Brazilian military, in conjunction with the American company Raytheon, has designed and is currently implementing in the region a satellite and radar surveillance network. The System for the Surveillance of Amazonia (*Sistema de Vigilância da Amazônia,* or SIVAM) aims at monitoring external threats to the national territory (illegal penetration by drug traffickers and others), improving air traffic control, and contributing to environmental protection (by identifying forest fires and invasions of protected areas).[43] The system suffered many delays, but in 1997 it began partial operations. Since the onset of Plan Colombia,[44] however, and due to the geopolitical tensions that it has generated, both SIVAM and *Calha Norte* have picked up steam and benefited from an increase in federal budget allocations for the year 2000.[45]

The same is true of infrastructure development plans for the Amazon region. The program *"Avança Brasil"* (Advance Brazil) was announced by the Cardoso administration in 2000 and should be implemented between 2002 and 2006. It consists of large-scale investments in infrastructure in the Amazon region. Among its goals are the duplication of existing highways, and the construction of ports, railways, and hydroelectric plants. While there is consensus on the need to improve infrastructure in the region, *"Avança Brasil"* has been criticized for its lack of concern with conducting environmental assessments of its initiatives and consulting with the Brazilian and Amazonian societies.[46]

This chapter has traced the evolution of development and environmental policies in Brazil, and in Amazonia in particular, during the past forty years. While not exhaustive, this summary revealed how local, national, and international processes have interacted to shape such policies. It also described the interests of the different actors that, operating in each of these three levels, have influenced efforts to promote the environmentally sustainable development of Amazonia. It is evident, though, that despite myriad initiatives, the goals of economic development and natural resources exploitation of Amazonia, and the region's environmental protection, remain far from being reconciled. It is under this basic constraint that the Rondônia transnational advocacy network operated, from its inception in the early 1980s until it demobilized at the turn of the millennium. In the following chapter I describe the origins of the network and its initial struggle against the Polonoroeste project.

3

Urgent Action! Transnational Mobilization against Disaster in Rondônia

Why and how did a group of individuals, and later of organizations, become interested in a remote corner of the Amazon forest in the early 1980s? What consequences did this interest have for the protection of the region's natural and human environments? What obstacles did activists encounter in their efforts to protect Amerindians and the environment, and how were they overcome? And finally, what impact did participation in a transnational environmental advocacy network have on Brazilian and Rondonian civil society groups? The history of the origins and evolution of the Rondônia transnational environmental advocacy network, or the Rondônia network, in the 1980s provides the answers to these questions.

In this chapter, I describe the formation of the Rondônia network and highlight the specific roles played by its main members at different junctures of the mobilization process. I also analyze members' relative political weight within the network, based on their different access to resources (political, material, and technical). Finally, I evaluate the network's strategies and choice of "targets" in light of what it accomplished—or failed to accomplish—during the initial years of its activism.

THE BEGINNING

World Bank officials were among the first individuals to raise concerns about the Rondonian environment, as the agency prepared to fund the Polonoroeste project. Bank internal documents clearly indicate the reluctance that predominated within some of its sectors about "funding a major highway construction project in a largely unknown, but ecologically sensitive area."[1]

Independent consultants, mostly American anthropologists, also wrote reports for the World Bank echoing the concerns of the agency's staff and further stressing the project's risks to Amerindian populations. The consultants' main worry was about the institutional weakness of the Brazilian Indian Agency and the government's lack of commitment to the protection of Amerindian lands.[2]

At the December 1980 meeting of the American Anthropological Association, Brazilian and American anthropologists had an opportunity to further discuss the risks of Polonoroeste.[3] Shortly after this meeting, a group of people who shared academic or professional interests in Brazilian Amerindian and environmental issues began to mobilize against the project.

Their first move was to obtain the institutional support of major environmental non-governmental organizations (NGOs) in the United States. This became possible by reaching out to individuals whose personal history was linked to environmental issues in Brazil, and who occupied key positions in environmental think tanks, namely the National Wildlife Federation (NWF) and the Environmental Defense Fund—currently Environmental Defense.[4] Ideals and principles, but also the perception of institutional opportunities that could be generated by mobilizing against a project such as Polonoroeste, influenced the participation of American environmental organizations. They perceived the study of Polonoroeste as a chance for NGOs to produce independent evaluations of the international development role of international financing institutions (IFIs).[5] In addition, activism on the Polonoroeste provided an opportunity for the strengthening of ties between Northern and Southern environmentalist groups. These alliances were an essential component of efforts by Northern NGOs to establish legitimacy when addressing development and environmental problems in the South.[6]

As I will discuss later in this chapter, one of the most successful strategies deployed by the Rondônia network occurred within the context of the Multilateral Development Banks (MDB) campaign. The campaign, staged by U.S. environmental and human rights organizations, aimed at pressuring MDBs to become accountable for the environmental and social consequences of their development initiatives.[7] Key strategies in this process were educating U.S. congressional representatives about the link between development funding and environmental degradation, and alerting the international public to the harmful consequences of development projects. The leaders of the campaign expected that the pressures of the concerned public and of the U.S. Congress on the MDBs would eventually lead to the reform of such institutions. As the Polonoroeste became the main case study of the campaign, and the object of many congressional hearings, the Rondônia network gained supporters among U.S. Congress representatives and their aides, and among the specialized media.

The participation of Brazilian individuals and civil society organizations in the Rondônia network occurred under the constraints imposed by military authoritarianism. By the early 1980s, the military had succeeded in either exterminating or keeping under tight control most civil society groups. A few anthropologists and technical personnel, mainly through their work for Polonoroeste's monitoring or implementing agencies, contributed to the network's attempts to mitigate the project's impacts. José Lutzenberger, a renowned Brazilian environmentalist, was at times the only Brazilian voice active in the mobilization. Finally, as the network evolved, it found partners in Rondônia. These consisted of a few individuals living in the state and who, for different reasons, maintained contact with environmental research and advocacy organizations in Brazil or abroad.

In sum, the original composition of the Rondônia network included U.S.-based environmental and human rights NGOs; a few Brazilian anthropologists and environmentally concerned individuals, who either worked independently or within Polonoroeste's implementing and monitoring agencies; World Bank environmental staff and consultants for Polonoroeste; U.S. Congress representatives and their aides; and selected environmental correspondents in the international media.

One should not assume, however, that these groups and individuals shared identical goals and priorities, had comparable capacities for resources mobilization, or used unified strategies to try to counteract Polonoroeste's harmful environmental consequences. The MDB campaign did provide some coherence to the network's strategies of pressuring the World Bank to increase monitoring of Polonoroeste, and to reevaluate its approach to development initiatives. Inside Brazil, on the contrary, the attempts to protect Rondônia's social and natural environments depended on random opportunities, the courage and skills of committed individuals, and the extent to which their loose contacts with transnational activists and organizations lent them much-needed resources.

The weakest link of the Rondônia network in its earlier years was precisely the relations between Brazilian and transnational members. Not surprisingly, difficulties in relations between domestic and international members help in explaining some of the network's shortcomings. While international environmental NGOs, their allies in the U.S. Congress, and environmental correspondents in the international media shared complementary goals of reforming the MDBs and raising public environmental awareness, the objectives of Brazilian activists were relegated to a peripheral position within the international mobilization effort. The subordination of the interests of Brazilian network members to those of the transnational members occurred in two different ways. First, the concerns of Brazilian activists with Polonoroeste were often "re-framed" to fit into the agenda of Northern groups. Second, Brazilians had little say in the network's choice of strategies to influence Polonoroeste. This disconnect between the priorities of Southern and Northern groups helps to explain the limited impact that the Rondônia network had in mitigating Polonoroeste's consequences for Amerindians and the local environment. The sections below evaluate this disconnect, as well as the opportunities found by network members to overcome it.

THE RONDÔNIA NETWORK GAINS MOMENTUM

The MDB campaign provided the strategic frame within which the Rondônia network gained momentum. Before knowing about the Polonoroeste project, Bruce Rich, a lawyer for the EDF had started to challenge the environmentally unsustainable lending practices of MDBs. Barbara Bramble, heading the international programs at the National Wildlife Federation, shared with Rich her knowledge about Amazonia's development problems and concerns with initiatives such as the Polonoroeste project. The two activists concluded that the Polonoroeste project was the perfect case study for the campaign. EDF and NWF soon obtained key resources to launch the campaign through an offer by Brent Blackwelder, then director of the Environmental Policy Institute, to lend his organization's lobbying structure.[8] It should be noted however that the campaign's most important asset in its initial phase was the political relevance of its leading organizations, in particular the NWF. Bramble deserves the credit for having convinced the organization to throw the weight of its more than five million affiliates behind the MDB campaign.[9]

In the early 1980s, the MDB campaign had two objectives. In the short term, activists attempted to interrupt World Bank funding for Polonoroeste. In the long term, the goal was to force multilateral development banks to promote structural reforms in order to address the environmental impacts of their lending policies.[10]

Between 1983 and 1984, the MDB campaign organized six congressional hearings about the environmental performance of multilateral development banks, and this number rose to seventeen by 1986.[11] The September 1984 Congressional hearing[18] was a landmark for the Rondônia network. It generated important opportunities for cooperation between U.S. environmental NGOs and Brazilian activists. The American groups, supported by data provided by Brazilian activists, presented high-quality testimonies about Polonoroeste's impacts that impressed congressional representatives. Rather than portraying themselves as spokespersons for Brazilian civil society, U.S. NGOs invited José Lutzenberger to share with the audience his view on the project. Finally, the hearing concluded with the presentation of a documentary film, *The Decade of Destruction,* produced by Adrian Caldwell, a British director, in partnership with staff from a Brazilian university. The documentary contained shocking scenes of forest fires and emaciated Amerindians left behind in the path of frontier expansion.

Through the hearings in the American Congress, the MDB campaign gained visibility. It benefited from media exposure,[13] gained the allegiance of NGOs throughout the world, and succeeded in involving environmental NGOs' affiliates by asking them to write letters to their representatives and to the World Bank's president, demanding the interruption of Polonoroeste. Yet, with the exception of an increased number of monitoring missions to the Polonoroeste area,[14] the World Bank seemed impervious to the campaign's pressures.

By late 1984, the campaign coordinators realized that an alliance with Democratic legislators in a Congress under Republican control was reducing their chances of success.[15] In a controversial move, the environmental coalition tried a bolder strategy. It turned to Republican senator Robert Kasten, from Wisconsin, for support.

Senator Kasten, at the time occupying the chairmanship of the Foreign Operations subcommittee of the Senate Appropriations Committee, was known as a fierce opponent to American international development aid. The World Bank's poor environmental performance provided a powerful argument to push for the reduction of U.S. contributions to the bank. This reduction would have severely threatened the World Bank's funding for its soft loans window, the International Development Association (IDA). Environmentalists were aware of the risks represented by this alliance. Their goal was not to push for the reduction of the bank's resources for development, but to redirect them to environmentally sound causes.[16] If properly managed, the threat to IDA's funding could become a powerful instrument for the achievement of the goals of the MDB campaign.

Not all of the sponsors on the MDB campaign agreed with the alliance with Republican sectors of the U.S. Congress.[17] Environmental and human rights NGOs on the far left of the American political spectrum rejected the move on principle. World Bank environmental staff, who until then had applauded and even aided the campaign by leaking information to campaigners and supporting their warnings in environmental journals, ceased to participate in the effort.

Despite its challenges, the alliance with Kasten was a turning point in the MDB campaign and, to a certain extent, in Polonoroeste's implementation. After the first hearing chaired by Kasten in June 1984, the issue of IDA's appropriations gained momentum within the Senate. It received even more attention in light of Kasten's reaction to a World Bank response to environmentalists' inquiries on Polonoroeste. On October 12, 1984, representatives of thirty-three organizations, among them major environmental groups in the United States, Europe, and in the developing world (Brazil, Ecuador, India), as well as members of the Brazilian Congress and the West German Bundestag sent a letter to the World Bank expressing concern with Polonoreste's implementation. The chief of Brazil's division in the World Bank's Latin America and Caribbean Regional Office answered the letter in a single paragraph, in which he affirmed the bank's close monitoring of Polonoroeste.[18]

Senator Kasten considered such a response "an insult." He immediately wrote to World Bank president Clausen demanding from him a "credible and responsive answer" to environmentalists' criticisms of Polonoroeste. Kasten's office also forwarded letters to the treasury secretary, Donald Reagan, and to the incoming secretary, James Baker, insisting on Treasury oversight of the issue.[19]

The World Bank's response to Kasten and to the Rondônia network arrived in less than twenty-four hours.[20] President Clausen not only sent a long letter to Kasten explaining recent negotiations with the Brazilian government on Polonoroeste's environmental problems, but also invited both the senator and the U.S. environmentalist organizations to a meeting at the World Bank, where he would provide further explanations. Clausen's letter to Kasten was dated March 1, 1985. On March 15, a World Bank letter to the Brazilian government stated that funding for Polonoroeste would be temporarily discontinued. The link between the pressures of the Rondônia network through the MDB campaign and the interruption of Polonoroeste is evident. For the first time in the history of the MDBs, the poor environmental performance of a project provided justification for its suspension.

While the "unholy" alliance between Senator Kasten and the environmentalists was a powerful leverage on the World Bank, it alone cannot explain the agency's unprecedented decision to suspend funds for Polonoroeste. The indignation of both the public and the international media about the environmental devastation caused by the project also affected that decision.

The success in mobilizing public opinion against Polonoroeste was, to a large extent, a function of a strategic selection of issues to be highlighted by the Rondônia network. Environmental problems caused by, or related to, Polonoroeste were several. They ranged from inadequate use of available land to the lack of sanitary infrastructure in overpopulated urban centers, from deforestation to soil degradation and an increased incidence of tropical diseases, and from the displacement of traditional populations (rubber tappers, fishermen), to the extermination of Amerindians and their cultures. For the layperson, the link between Polonoroeste and many of these issues required detailed and complex explanations. There was, however, a subset of environmental problems caused by Polonoroeste that could capture the imagination of lay audiences throughout the world. It included the destruction of tropical forests and the plight of Amerindian groups in early stages of contact with the Brazilian society. For the American environmental and human rights NGOs, the selection of such a subset of issues as the network's priority was self-evident.

Brazilian activists, however, resented this somewhat "romantic" and ethnocentric approach to environmental and development problems in Amazonia.[21] For them, among Polonoroeste's most serious issues, were its neglect of ecosystems such as the *cerrado*,[22] of crucial importance for biodiversity, as well as the project's failure to assist populations who were the key vectors of deforestation, that is, migrants and small farmers. A Brazilian environmentalist highlights the implications of the differences in Northern and Southern approaches to Polonoroeste's environmental problems:

> There is a significant divergence between agendas. American and European eyes are turned toward forests, and they do not notice many other things. For instance, the *"cerrado"* has been destroyed by almost fifty percent. But who speaks for the *"cerrado"*? It is the area of savannah with the highest degree of biodiversity in the world. Yet, once again, the imaginary remains more important than what is rational. The romantic approach to the forest and to Amerindians, which is disseminated abroad, is very different from what really exists here [in Brazil]. The indigenous question represents an enormous diversity—there are those who desire to remain isolated and those who have been selling all that they can from their lands in order to maintain a certain quality of life. It is therefore a very complex question that cannot be limited to the marketing process. As we know, marketing strategies always choose the shortest way. But this generates terrible distortions that affect the understanding of the problem.[23]

Part of the difficulty in transforming many of the environmental problems caused by Polonoroeste into campaign flags was the insufficient information available. In the early 1980s, the impact of human interventions in Amazonian ecosystems was still poorly known.[24] Many seminal studies establishing links between Brazilian preexistent political and economic structures, and environmental problems were conducted mostly during or after Polonoroeste's implementation.[25] On the other hand, knowledge on Amerindians and on the biodiversity value of tropical forests had been available in academic and environmental communities since the late 1960s.[26] Anthropologists such as Claude Lévi-Strauss, for instance, had made the Nambikwara indigenous group, a population greatly affected by Polonoroeste, known in anthropological circles worldwide.

The second reason for focusing on tropical deforestation and Amerindians was their appeal to the international media.[27] Miles of tropical forests burning for months in the dry season as well as emaciated Amerindians suffering from diseases long eradicated in the industrial world, instantaneously provoked the public's rage and, simultaneously, made powerful newspaper and television headlines.

The media's interest in the Polonoroeste project greatly added international visibility to the MDB campaign and increased its leverage over the World Bank. In Brazil, however, the interest of the international media in Polonoroeste and the international public's outrage against the project contributed little to the initiatives of the members of the Rondônia network.[28] To a certain extent, Brazilian political conditions contributed to this situation, since the government censored open criticism of its policies. In addition, the level of environmental awareness of the Brazilian public in the 1980s was still much lower than that found among international audiences during the same period.[29] How, then, did Brazilian and transnational members of the Rondônia network attempt to affect Polonoroeste nationally, and most important, locally?

Transnational Activism and Domestic Constraints— The Rondônia Network within Brazil

The political and social constraints on environmental activism in Brazil in the early 1980s forced members of the Rondônia network to resort to strategies more informal, and less structured, than those used within the scope of the MDB campaign. Environmentally committed individuals within the World Bank's Polonoroeste team[30] were the first among those who attempted to influence the design of the project. Their concern with the impact of the project on Amerindian populations and on the region's ecosystem, when added to other expectations inside the bank regarding Polonoroeste, led to project redesign.[31] Rather than simply a road project, as it was originally proposed by Brazil, Polonoroeste became an integrated development program, including agricultural settlement, environmental and Amerindian protection, and research components.[32] Brazilian agencies in charge of these issues at both local and federal levels were thus brought together to implement the project.

The redesign of Polonoroeste created two important areas through which concerned individuals and organizations could influence the project. The first was the creation of an Amerindian Special Project (ASP). The second was the project's monitoring component, under the coordination of a team from the Institute of Economic Research Foundation (FIPE). I discuss below the activism of the Rondônia network in each of these arenas as well as the obstacles that it encountered.

It was the World Bank, at the recommendation of some of its staff, that insisted that the Brazilian Indian Agency, FUNAI, draft a sub-project to protect the physical and cultural integrity of Amerindians. Protection, rather than integration of Amerindians, contradicted the dominant approach to the "Indian question" in Brazil. The military's national security ideology opposed indigenous peoples "protection" and autonomy. It favored these populations' integration into Brazilian society, which implied the complete abandonment of their "primitive" ways.[33] In spite of its resistance, FUNAI eventually rushed to draft the Polonoroeste's Amerindian Special Project, and did so only to meet the Polonoroeste loan agreement's conditions and deadlines.[34] Yet, to clearly indicate Brazil's rejection of external interference on Amerindian issues, the government refused direct bank funding for ASP.

The story of the implementation of the Amerindian Special Project highlights the importance of individuals and of their personal commitment to environmental and human rights issues in transnational advocacy networks. There were two very distinct phases in the process of implementing the Amerindian project. In the first, between 1983 and 1985, individuals committed to Amerindian protection were in influential positions both within the World Bank Polonoroeste team and at the Brazilian Indian Agency. Within this context, the work of Brazilian anthropologists consulting for the project received institutional support, and their proposals on behalf of Amerindians were satisfactorily implemented.[35] Support from World Bank staff to the implementation of Amerindian protection measures occurred in the form of meetings with Brazilian agencies to demand information on the status of demarcations and to expedite the allocation of funds for specific initiatives. Consultant anthropologists supported these actions and provided technical information to World Bank staff through both formal (consultant reports) and informal (private meetings and personal letters) channels. This informal alliance between World Bank staff, selected FUNAI technical personnel, and consultant anthropologists was probably the only engine behind the accomplishment of a limited number of Amerindian protection measures within Polonoroeste (namely, vaccination campaigns and land demarcations).

Between 1985 and the completion of the project in 1987, this informal alliance was broken. An atmosphere of antagonism and confrontation against the work of consultants and committed staff at FUNAI on the part of World Bank managers and Brazilian officials replaced the previous cooperative climate and reversed most of the small conquests of the pre-1985 period. The rise of politically conservative individuals to the presidency of FUNAI and the leadership of Polonoroeste (in the role of project task manager) was a key factor in this process.

The changes in the Polonoroeste World Bank team also affected the project's monitoring component. FIPE, an independent academic institution, was in charge of project monitoring, particularly of its environmental and Amerindian components. Relations between FIPE and Polonoroeste's implementing institutions were often tense.[36] It was only through the mediation of World Bank staff that the FIPE team was able to continue with its work until 1987. After that year, however, FIPE's contract was not renewed. Brazilian consultants and even World Bank's internal documents attribute the decision of relieving FIPE from its monitoring role to the change in personnel within the World Bank's Polonoroeste management team.[37] Selected segments of interviews illustrate the importance of the collaboration between World Bank staff, Brazilian consultants, and technical personnel in Brazilian implementing agencies for the implementation of some of Polonoroeste's environmental and Amerindian conditions (it also highlights the difficulties that committed Brazilian individuals faced during the project's final years):

> [Bank staff] organized meetings in which they demanded results. The ASP contract had committed the Brazilian government to a series of actions. [Bank staff] then asked: "What has been done?" "Have Amerindian lands been demarcated?" "If not, where is the money allocated for it?" . . . [Bank staff] could then show disapproval of FUNAI's president based on the fact that ASP's commitments had not been met. Sometimes, we [Brazilian anthropologists] participated in such meetings, sometimes we met independently. We had meetings with Amerindians to discuss their situation, and with government officials to discuss Amerindians' problems. Often times leaders of Amerindian populations came with us to such meetings. At that time, indigenous associations were starting to form. On one occasion, I had a meeting with the FUNAI president and with eighty Amerindian leaders![38]

> FUNAI's general tendency is contrary to Amerindians' interests. But there were always individuals there who did not share that approach. For instance, there was once a staff member, who almost became the agency's president, who directed the agency's lands division. He was wonderful! He was successful in demarcating lands by going over one thousand bureaucratic requests, most of them illegal. . . . The FUNAI's land team was great! They were real allies. . . . But within FUNAI there was a fierce struggle. On one side there were some people really committed to a program for the protection of Amerindians—and their struggle was heroic—and on the other, there was everything else that opposed this goal.[39]

> (When the FUNAI administration changed), it got to the point of forbidding consultant anthropologists from entering Amerindian areas. We faced all sorts of problems. That was because the anthropologists would visit the areas and often verified that local FUNAI personnel were involved in illegal activities. This was denounced in our [monitoring] reports.[40]

[After the change in Polonoroeste management] FIPE suffered many pressures due to its commitment with the implementation of Amerindian and environmental conditions. At one point, the Brazilian government wanted one of our reports altered. FIPE's president supported the consultants and refused to change their conclusions. Shortly after that our monitoring contract was not renewed.[41]

TAKING STOCK OF THE RONDÔNIA NETWORK IN THE 1980S

Both internal and external factors affected the Rondônia network and its capacity to mitigate the impact of the Polonoroeste project on Rondônia's natural and human environments. Internal factors related to structural differences between network members and the cleavages that sometimes set them apart. External factors referred to Rondônia's specific political and socioeconomic realities. These realities posed challenges to the network members that they were not prepared to meet.

The Rondônia network, at its inception, was characterized by striking differences among the relative capacities of its members. Their access to financial, technical, and political resources varied greatly. As a consequence, there was an inevitable predominance of certain members over others regarding the selection of the network's priorities, strategies, and targets. Different approaches to these issues sometimes generated divisions within the network that hindered its ability to achieve important goals.

The technical expertise of World Bank environmental staff and the amount of financial resources (in the form of trips to the field, communication facilities, and general infrastructure) available to them were major resources for their role in catalyzing environmental and Amerindian-related initiatives in Brazil. Yet bank staff's major resource was institutional. Since they were seen as "World Bank people," and thus personified the institution's political and economic might, selected bank staff were able to influence Brazilian governmental agencies and, on occasions, support dissident voices such as those of Brazilian consultants.

As for the Washington-based NGOs, their leadership role within the network was derived from their legitimacy and political weight within American civil society. The National Wildlife Federation, as stated above, was at that time the largest conservation organization in the United States and in the Western world, with five million members and supporters, while the Environmental Defense Fund had more than 100,000 affiliates in the United States.[42] By themselves, these organizations' constituencies gave them the political leverage to influence the U.S. Congress and the World Bank itself. They certainly added to the financial resources available to the MDB campaign.

Yet, to affect issues in a foreign country, international NGOs needed more powerful leverage than their number of affiliates. This leverage came by devising arguments that justified foreign concerns in domestic terms. In other words, because American and European taxpayers' money funded the institutions that financed projects in the developing world (i.e., MDBs), these institutions had to be accountable to taxpayers' interests. As such, the argument went, MDBs had to see that their loans were not contributing to results (environmental degradation, for instance) that were opposed by their primary sponsors.

As coherent as this "accountability argument" may have been, it was not sufficient to completely eliminate tensions related to sovereignty issues among network members. An alliance with Northern groups who were critical of Brazilian policies made Brazilian activists vulnerable to charges of cooperating with international imperialism. As one consultant explains: "[W]hen pressures led to the suspension of funds for Polonoroeste, due to problems in Amerindian areas, we [from FIPE] were accused of collaborating with an external institution and against Brazil. . . ."[43] Inevitably, some Brazilian members became weary of open collaboration with international groups.

The participation of Brazilian activists in the MDB campaign was somewhat tainted by concerns with issues of national sovereignty. Yet the most serious obstacle to such participation was the inequality of material and political resources between Northern and Southern groups. Sometimes consultants for Polonoreste felt that they had not been given the credit they deserved for providing the MDB campaign with important information on the project.[44] Whereas part of the problem was the predominance of international NGOs as the campaign interlocutors, Brazilians recognized that their own lack of financial resources, for traveling abroad, for instance, hindered their role.

The disproportionate amount of resources invested in the MDB campaign also generated tensions within the Rondônia network. As I have discussed above, network members held different views on the most relevant environmental issues in Rondônia and on how to address them. Northern groups emphasized the plight of Amerindians and the devastation of tropical forests, and linked these outcomes to the unsustainable development model sponsored by MDBs. According to this perspective, the solution to environmental problems in Amazonia, and elsewhere in the developing world, was to reform the lending policies of such institutions. The MDB Campaign thus was the primary instrument for achieving this goal.

For the Brazilian network members, the "greening" of MDBs was a distant priority. Their primary concern was with the Brazilian military's authoritarian policies of occupation of Amazonia. The ultimate goal of Brazilian activists was to pressure the government to change such policies, while protecting Amazonian populations—Amerindians and others—against their detrimental socioenvironmental effects. The MDB campaign was one instrument in this struggle, but it was neither the only one nor the most important. In retrospect, the emphasis of the most resourceful members of the Rondônia network on the campaign limited opportunities for formulation of alternative strategies that could more appropriately address the priorities of national groups. For instance, the Rondônia network was not able to devise a single strategy to support the informal alliances between FUNAI personnel and Brazilian activists that led to the few accomplishments of Polonoroeste's early years regarding Amerindian protection.

While differences in members' relative capacity and selection of priorities constituted internal difficulties that hindered the effectiveness of the Rondônia network, the network's major challenges were the political and socioeconomic dynamics in the Polonoroeste region. Mostly due to insufficient knowledge available on the Brazilian Amazonia's northwest region, the members of the Rondônia network were unable to fully assess the intensity of processes of frontier expansion. As a consequence, they lacked the capacity to devise strategies that could most effectively reduce the environmental impacts of these processes.

Throughout the 1980s, the members of the Rondônia network focused their attempts to mitigate Polonoroeste's impacts on two main actors—the World Bank, mainly through the MDB campaign, and the Brazilian federal environmental and Amerindian agencies. The assumption supporting the selection of such "targets" was that these institutions were, in fact, capable of implementing the project's socioenvironmental conditions. Activists identified these institutions' lack of political will as the main obstacle to the fulfillment of their commitments. They expected that political pressures in international arenas would alter this situation.

In time, what became evident was that neither the World Bank, nor the Brazilian federal government, let alone Brazilian environmental and Amerindian agencies, had the capacity to control the forces and processes leading to environmental devastation in Rondônia. In hindsight, the most prudent course of action (often recommended by environmental NGOs and independent consultants) would have been to either interrupt or slow down disbursements for the project. Instead, the Polonoroeste project was characterized by the fast pace of its loan disbursements and implementation of infrastructure components (namely, the paving of the highway). Thus, rather than controlling processes of frontier occupation that had been initiated in Rondônia in the mid-1960s, the project contributed to accelerate them.

Without going into further detail, it is important to summarize the political and socioeconomic dynamics that shaped the occupation of the Brazilian northwest frontier, and of Rondônia in particular, during the 1980s. These dynamics continued to challenge the efforts of the Rondônia network to protect the state's socioeconomic environment throughout the 1990s.

The "boom" of logging operations in Rondônia coincided with the beginning of Polonoroeste implementation (1980–1984). The wood industry was greatly favored by Brazil's policy of subsidizing credit for export activities. Polonoroeste only made wood extraction and transportation easier by improving the conditions of highways and feeder roads.[45] During the implementation of the project, the logging industry in Rondônia grew from 400 registered operations in 1982, to 1,150 in 1987.[46] As logging intensified and high-quality wood became scarce in public lands, extraction work moved into Amerindian lands. Often, logging companies had permits issued by FUNAI itself.[47]

Large deposits of bauxite, zinc, gold, and cassiterite had attracted both large-scale mining enterprises and small prospectors to Rondônia since the 1970s. From 1980 to 1986, the Brazilian National Department of Mineral Research (DNPM) approved 164 claims for mineral research and exploration within indigenous lands and protected areas in the state. Parastatal companies, private national groups, and multinationals were among the recipients of these concessions.[48]

The expansion of both mining and logging operations in the Polonoroeste region are consequences of a combination of availability of abundant natural resources and governmental incentives to frontier occupation. While these processes have had significant impacts on the region's environment, they hardly compare to the effects of accelerated migration to the area.[49] Population numbers for Rondônia show that from 1980 to 1991 the state's resident population increased 7.88 percent, whereas the national rate was 1.93 percent.[50] Despite Polonoroeste's attempt to regulate migration and establish coherent settlement policies, it was unable to respond to increasing demands for land and infrastructure. Even migrants that came to Rondônia under the sponsorship of the project remained unassisted. Once implementing agencies failed to provide technical and social support for small farmers new to the region, their subsistence became almost impossible. Given the unexpected low agricultural potential of many lots,[51] slash-and-burn practices grew rampant as settlers produced for their immediate needs.[52] After a few years, soils had been exhausted and the plots abandoned or sold for a minimum price. Deforestation, invasion of Amerindians' land, and land concentration were among the main consequences of these processes.[53]

Land concentration and the expansion of cattle-ranching activities were inevitable consequences of the failure of Polonoroeste settlement policies.[54] Unable to survive on their small plots without government assistance, small settlers sold them for very little or simply abandoned them.[55] By then, unsustainable agricultural practices had already exhausted the soil and cattle ranching was the most economically rational option.[56] Cattle ranching required almost no investment in infrastructure and labor, and benefited from tax breaks and subsidized credit. As explained in the previous chapter, these incentives were granted almost automatically, against proof of land "productivity," that is, deforestation.[57]

Finally, local political and electoral interests have had a direct impact on Rondônia's human and natural environments. Previously a federal territory, Rondônia achieved statehood in 1982, one year after the beginning of Polonoroeste implementation. To enhance its political weight at the federal level, the new state strongly encouraged migration (as in the United States, the larger a state's population, the higher the number of its representatives in the country's house of representatives). Locally, patron-client relations, a political practice by which local politicians distribute state resources in return for political support and votes, have dominated politics. This practice had significant impact on the intensification of land conflicts and deterioration of environmental conditions.[58] For instance, it was common, and still is, for a Rondonian politician to channel state resources toward settlement projects in forest reserves and even on Amerindian lands.[59] Settlers' support translates into votes, while Amerindians did not vote until after the 1988 constitution.

In light of such powerful political and economic obstacles to environmental protection, what did the efforts of the Rondônia network members accomplish in the 1980s? The answer varies, depending on one's priorities.

At the international level, analysts are unanimous in giving credit to the MDB campaign, and to its activism against the Polonoroeste project in particular, for its accomplishments in reforming the World Bank (and to a lesser extent, the other multilateral development institutions).[60] While environmentalists continue to criticize the bank for many of its development policies, it is evident that the institution has reformed to a degree that currently prevents projects, such as Polonoroeste, from even being considered.[61] The activism around Polonoroeste was also important in setting a precedent that sent shock waves across the international development community: environmental concerns could lead to the interruption of international funding to development initiatives. Finally, the campaign had consequences for Brazilian environmental policies. Its ability to mobilize international public opinion against Amazonia's deforestation eventually contributed to encouraging internationally funded initiatives to protect the region (namely, the Pilot Plan and the National Program for the Environment, PNMA), as described in the previous chapter.

The Rondônia network was unable to affect Brazilian environmental and Amerindian agencies in the long term. In the short term, particularly between 1983 and 1985, an alliance between World Bank staff, consultant anthropologists, and FUNAI's committed personnel strengthened the latter. As a result, there were small gains in the areas of Amerindian health and land demarcation.[62] These gains were compromised in later stages of project implementation, since they depended, entirely, on the personal commitment of selected individuals.

At the local level the Rondônia network was unable to devise strategies to either channel resources or affect political will toward environmental and Amerindian protection. Two factors explain this outcome. First, in the early 1980s there was an almost absolute lack of knowledge on environmental conditions in the Amazon frontier, and on how they were impacted by accelerated development. The Rondônia network deserves credit not only for sponsoring research on these issues, but also for heightening academic interest in them. The second and most important factor affecting the network's effectiveness at the local level was the absence of local NGOs and grassroots groups in the mobilization effort.

This structural weakness of the Rondônia network made it impossible for its members to effectively monitor Polonoroeste's initiatives and mitigate local processes of environmental degradation that the project accentuated. Even the Brazilian members of the network did not reside in the area and did not even control the resources that allowed them to travel there (mostly funds from Polonoroeste's monitoring component). The few individuals who lived in Rondônia and were committed to mitigating the impact of Polonoroeste eventually produced important knowledge on the area's environmental problems. Alone, however, they did not have either political leverage or material resources to address them. Both national and international members of the Rondônia network became aware of this structural weakness in the mid-1980s. Whereas at the beginning of the Rondônia network's activism one could hardly find a single local organization that spoke for the interests of Amerindian groups, migrants, and those concerned with the environment, by the end of the decade, many international and Brazilian NGOs, as well as national social movements had dedicated resources to the establishment of local chapters in Rondônia. As the reader will observe in the next chapter, this process had important consequences both for the network itself, and for Rondônia's environmentally sustainable development.

4

"Localizing" Transnational Activism— Success and Failure

Despite the completion of the Polonoroeste project in 1987, the Rondônia network remained mobilized. The main reason for continued concern with the fate of the area's human and natural environments was the imminent prospect of a "Polonoroeste II." Even before the formal completion of the Polonoroeste project, state and federal government officials and World Bank staff had started to discuss funding schemes for further development initiatives in Rondônia.[1] For some individuals within this constituency, the new project would offer an opportunity to correct the distortions that Polonoroeste had caused.

The process of formulating "Polonoroeste II," or as it was later named, the Rondônia Natural Resources Management project (Planafloro, for its acronym in Portuguese), lasted from 1987 to 1992. On March 17, 1992, it was finally approved by the World Bank board of directors and in September 1993 its implementation began. There were two reasons for such a long and cumbersome process of preparing Planafloro. The first was the reluctance of both the Brazilian and Rondonian governments to request external funding for the state. Given their experience with Polonoroeste, both governments were wary of the inevitable scrutiny of their actions that a new development initiative in the area would generate. Furthermore, the federal government did not want to endorse further borrowing from an insolvent state such as Rondônia. The second reason for delays was the close monitoring of members of the Rondônia network over Planafloro negotiations. In this chapter I discuss the relative success of the network in affecting Planafloro's design, particularly by including in it strict environmental preconditions to the loan agreement as well as mechanisms for local civil society participation. I also address the shortcomings of the

Rondônia network, which accounted for its legitimacy crisis in 1994. The story of the Rondônia network in the early 1990s illustrates the challenges of "internalizing" or "localizing" transnational environmental activism.

At the time that the Planafloro negotiations started, the membership of the Rondônia network had grown and diversified significantly vis-à-vis its original composition in the early 1980s. International environmental and human rights NGOs involved in the MDB Campaign had established an informal "division of labor." Among other things, that meant that NGOs with a comparative advantage in certain regions or countries would be primarily in charge of networking with that region's or country's national and local groups, as well as coordinating strategies within—and outside—the MDB Campaign's context. In the case of Brazil, and particularly of Rondônia, this role fell on the EDF, and to a lesser extent, on Oxfam, UK. As Planafloro implementation evolved, Friends of the Earth, Amazonia Program (FoE), and the Center for International Environmental Law (CIEL) became the main international partners in the network.

In Brazil, the democratization process and increased interest in environmental issues generated political and financial opportunities for the emergence of environmental NGOs and research institutes. A group of consultant anthropologists for Polonoroeste, for instance, founded the Institute of Anthropology and the Environment (*Instituto de Antropologia e Meio Ambiente*—IAMA). Although IAMA's headquarters were in southeast Brazil (São Paulo), it worked closely with a small partner organization in Rondônia,[2] and with local activists. The Institute for Amazonian and Environmental Studies—became another important member of the Rondônia network at the national level. Different from IAMA, IEA did not have a history of participation in the mobilization against Polonoroeste. Its interest in Rondônia, and in Planafloro in particular, derived from its work with the then recently formed National Council of Rubber Tappers (*Conselho Nacional dos Seringueiros*—CNS). Rondonian rubber tappers were among the main beneficiary populations of Planafloro,[3] yet until 1991 they had not succeeded in creating an organization that represented their interests. Although IEA's headquarters were initially in Curitiba, in the south of Brazil, and later in Brasilia, the country's capital, the institute opened a small office in Rondônia in 1992. The partnership between IEA and the rubber tappers' council brought to the Rondônia network the collaboration of a key activist, Chico Mendes, whose role I describe below.

In Rondônia, the number of individuals and organizations concerned with environmental issues had increased since the early 1980s. In part, the devastation caused by Polonoroeste contributed to such an outcome. Technical personnel who were once involved in implementing Polonoroeste had grown weary of working for the government and moved to the private sector.[4]

The establishment of local chapters of environmental and human rights groups was a response to pressures from international funding agencies, who required a locally based entity as the recipient of grants and project funds.[5] To a large extent, local activists staffed these Rondonian chapters. Church-based organizations that had operated in Rondônia since the 1970s, such as the Catholic Church's Indigenous Peoples' Missionary Council (*Conselho Indígena Missionário*—CIMI) and the Pastoral Commission on Land (*Comissão Pastoral da Terra*—CPT) became increasingly aware of the links between major environmental devastation and the declining quality of life of their constituencies. Finally, grassroots groups at very early stages of organization, such as the Karitiana Indigenous Peoples's Organization, the Union of Indigenous Peoples of Rondônia (*União dos Povos Indígenas*—UNI-RO), the Organization of Rondonian Rubber Tappers and the Rondonian chapters of the rural workers' union and of the Landless People's Movement (*Movimento dos Trabalhadores Rurais Sem Terra*—MST), began to raise questions about Planafloro.

But why did Planafloro raise so many concerns, despite the fact that its stated goals were to improve management of Rondônia's natural resources, and how did the Rondônia network address such concerns? The answers to these questions relate to two main issues: the lack of transparency in the project design process and the lack of civil society participation in project implementation and monitoring. Thus, as official negotiations on the Planafloro project unfolded, the Rondônia network pursued two distinct yet complementary goals in response to the concerns of its members. The first goal was to increase project transparency by forcing official negotiators to consult with affected communities and more closely address their expectations. The second was to expedite capacity building among grassroots groups and environmental and human rights NGOs in Rondônia in anticipation of their probable role in Planafloro implementation. As I discuss these goals in the sections below, I highlight the opportunities and challenges that their pursuit represented for local groups and for the structure and effectiveness of the network itself.

The Issue of Transparency: Evironmentally Sustainable Development To Whom? By Whom?

Planafloro's original formulation by the government of Rondônia made it essentially a development project. The project's title in Portuguese translates as Agro-livestock and Forestry Plan of Rondônia.[6] But if Rondonian elites essentially wanted World Bank money for continuing their "business as usual practices" of exploiting the state's natural resources, the bank had different expectations.[7] Planafloro represented an opportunity to respond to critics, particular those leading the MDB campaign, and to bolster the institution's image as a champion of environmentally sound development. Lending further visibility to such a prospect was the possibility that the formalization of a loan agreement for Planafloro would coincide with the 1992 United Nations Conference on Environment and Development, in Rio de Janeiro, Brazil. The interest of bank officials in funding a "green" project eventually forced the hand of Rondonian officials, who agreed on expanding the scope of environmental and Amerindian protection components. The final version of Planafloro, however, was also significantly influenced by the activism of the Rondônia network itself.

The network's activism influenced Planafloro design in two ways: first, by strengthening a growing trend, within the World Bank, of formally including civil society participation in its projects; second by endorsing, albeit inadvertently, a "conservationist" or "mainstream" approach to environmentally sustainable development.[8]

The involvement of Chico Mendes in the MDB campaign, and to a lesser extent of other Brazilian grassroots activists, was an effective strategy to bring a "local face" to international arenas in which projects such as Planafloro were discussed. In the late 1980s, Chico Mendes was the president of the Rural Worker's Union and the leader of the rubber tappers of Acre, a neighboring state to Rondônia. His struggle against the destruction of the Amazon forest, for which he received awards from the United Nations Environmental Program (UNEP) and the Better World Society, had become known worldwide.[9] Mendes visited Washington, D.C., in both 1987 and 1988, when he met with MDB officials, U.S. Congress representatives, and high-ranking government employees to discuss the risks of development initiatives in Amazonia, and the rubber tappers' proposal for the creation of extractive reserves.[10]

Mendes issued the first warnings against Planafloro in an October 1988 letter to the bank's President, Barber Conable. He firmly stated that the government of Rondônia had no intention of creating the extractive reserves that it would commit to under the Planafloro agreement. On the contrary, project funds would finance colonization projects in pristine forests.[11] Chico Mendes never received a response from the bank. In December of that same year he was assassinated in his home state.[12]

The reader should note though that despite the intention of MDB campaigners to more closely involve local voices, Mendes was not a native of Rondônia. His interest in Planafloro and in populations potentially affected by it was indirect. It related to attempts to establish a Rondonian chapter of the national rubber tappers' council, CNS. Ironically, Rondonian rubber tappers resisted joining the council. In part, the competition that emerged between Brazilian environmental NGOs, namely, IEA and IAMA, over a preferential access and advisory role to rubber tappers, explains this resistance.[13] While IEA had established itself nationally and internationally as the primary partner of the council, in Rondônia IAMA was the one organization that had established historical links with rubber tappers and Indigenous peoples from the onset of Polonoroeste. Competition between support organizations and a history of tensions between Acre's and Rondônia's rubber tappers eventually led the latter to create their own local organization, the OSR, which is independent from the national council. At the onset of debates over Planafloro, both the council (still without a local base) and the OSR participated, but the council left Rondônia in 1993.[14]

The efforts to bring local voices to Planafloro negotiations did not end with Mendes's death. In 1990, Stephen Schwartzman, from the EDF, found out, almost by chance, that despite the World Bank's claims of consultation with local groups, most of Planafloro's potential beneficiaries had not been contacted.[15] EDF sent a letter to the bank with signatures from representatives of thirty-five environmental and human rights groups from Brazil, the United States, and ten other countries, in which they expressed concerns with the way project negotiations were conducted. Activists' main charges were, first, that contrary to the bank's claims, project managers had not consulted with local NGOs, professional organizations, unions, or community groups about Planafloro. A second concern was that the project did not contain any provisions for NGOs' participation in its implementation. Finally, the EDF letter listed a series of environmental measures that Polonoroeste had failed to implement, and questioned the bank on the wisdom of funding Planafloro when its predecessor remained unfinished.[16]

While the bank discussed internally how to best reply to the EDF letter, it received correspondence from the Brazilian minister of environment, José Lutzenberger. The unexpected nomination of the Brazilian environmentalist to head the environmental ministry had been a part of a strategy by President Collor's administration to bolster its environmental image. Both the Brazilian and international environmental communities welcomed the decision. For the Rondônia network, in particular, the nomination represented a valuable resource, due to Lutzenberger's history of cooperation with the network's initiatives. Activists did not lose time in requesting that the minister endorse their concerns about Planafloro. In March 1990, Lutzenberger sent a letter to the bank supporting the demands of the Rondônia network members for "a broad consultation and greater participation of non-governmental organizations in [Planafloro]."[17] It stated that consultation had not occurred at satisfactory levels and requested "substantial changes in the project."

The pressures of the Rondônia network, combined with Brazil's acute financial crisis in 1990, led to the temporary withdrawal of Planafloro from the bank's lending agenda for that year. Negotiations on the project, however, continued and in 1992 the bank's board of directors approved a revised version of Planafloro.

The Rondônia network did not endorse the bank's approval of Planafloro in 1992 any more than it had two years later. Yet the reasons for the network's opposition were different on each occasion. In 1992, members of the Rondônia network acknowledged that, at the level of design, Planafloro represented a step forward in promoting environmentally sustainable development in Rondônia. They endorsed the project's key environmental task, which was to develop in-depth studies of Rondônia's soil and natural resources in order to rationalize the state's occupation. The main tool in this process was the formulation of a socioeconomic and environmental zoning plan for Rondônia. Environmentalists, however, were extremely critical of the fact that both the bank and the Rondonian state had completely ignored the environmental preconditions to the project that the network had pressured for and that eventually became part of the loan agreement. While these were measures that the Rondonian government had to implement before Planafloro was signed, the project went forward without the completion of a single precondition. I will discuss the story of the Rondônia network's struggle for the implementation of Planafloro's environmental preconditions in the following section. For now, it is important to explain how the preconditions reflected the "conservationist" or "mainstream" approach to environmentally sustainable development that predominated within the Rondônia network in the early 1990s. (In the following chapter I will discuss how this approach evolved as a function of the growing influence of Rondonian groups within the network.)

It is important to recall that in the early 1990s, despite the growth and diversification of the Rondônia network's membership, its main catalysts remained the large international environmental NGOs. As had happened in the mobilization against Polonoroeste, the focus of international groups was on the creation and protection of conservation units in Rondônia, and demarcation of Amerindian lands. The demarcation of rubber tappers' extractive reserves also had become part of this subset of environmental priorities. This is clearly illustrated by the January 1990 EDF letter to the World Bank. In it, environmentalists demanded the completion of demarcations, expulsion of invaders, and protection of seventeen Amerindian reserves and four protected natural areas that should have been implemented under the Polonoroeste loan. They also required that the state guarantee the physical integrity of areas designated by the zoning plan for extractive activities, and implement measures to address the environmental and health problems caused by illegal gold mining along the Madeira river and its tributaries. A year later, this same subset of concerns continued to be the network's main goals and the object of a formal agreement between local network groups and the Rondônia government (see details of this agreement in the next section).[18]

Clearly, the Rondônia network continued to place great emphasis on environmental issues *"strictus sensus."* Yet the lessons of Polonoroeste, and the voices of Brazilian environmentalists, who were, to a large extent, more sensitive to the social and economic aspects of environmental degradation than their international counterparts, had started to influence the network.[19] Both the EDF letter and the 1991 agreement of local groups with the Rondonian state included demands that addressed social and economic issues, such as the state's technical and financial support to small farmers and its commitment to fund research on appropriate perennial crops for the region. The inclusion of socioeconomic demands within the scope of environmental activism eventually contributed to strengthening the interest and participation of important local grassroots groups in the network, namely, the Rondonian rural workers. At the same time, the fact that national and international members of the network easily embraced such demands suggests, among other things, the beginning of a shift in the level of influence that local groups (rural workers and others) exerted within the network. In the following section I describe the efforts of the Rondônia network toward capacity building among local organizations, and discuss how these efforts affected both the Planafloro project and the network itself.

BUILDING CAPACITY FOR PARTICIPATION: SEARCHING FOR THE RIGHT RECIPE IN RONDÔNIA

The recipe for building local capacity and encouraging local participation in public policymaking processes in Rondônia consisted of two trends. The first was a tendency by local groups to emulate the strategies of the MDB campaign and to operate within its framework; the second was a tendency toward institutionalizing participation, both within Planafloro and in the larger context of Rondônia's politics. A brief history of Rondonian civil society participation in Planafloro design and early stages of implementation illustrates these processes.

It would have been inconsistent for the Rondônia network to demand from Planafloro's negotiators that local groups be consulted about the project without simultaneously contributing to these groups' organization and capacity-building efforts. This was precisely what NGOs such as the National Wildlife Federation, Oxfam, and EDF did, in partnership with the Institute for Amazonian and Environmental Studies (IEA) and the National Rubber-Tappers' Council (CNS). In November 1990, these groups organized and financed a meeting among local NGOs and grassroots organizations to discuss Planafloro. The meeting counted on the political and logistical support of local chapters of the National Indigenous Union (UNI) and the rural chapter of the workers' confederation in Rondônia (CUT-rural). It was held in the state's capital, Porto Velho, and became a landmark for coordinated actions among Rondonian groups concerned with environmental and human rights issues.

Shortly after this meeting, by mid-1991, many Rondonian NGOs and some grassroots groups created the Forum of NGOs and Social Movements of Rondônia *(Forum de ONGs e Movimentos Sociais de Rondônia)*, from here on, the Rondônia Forum. The forum's design originated during discussions between IEA's local and national staff and members of international NGOs.[20] It was conceived as an umbrella organization, with the mandate of facilitating civil society participation and monitoring environmental and development policies in Rondônia. Funding for the forum came mainly from Oxfam, and later from the World Wildlife Fund for Nature (WWF).

Almost simultaneously with the creation of the Rondônia Forum, local civil society groups, under the leadership of IEA's local staff,[21] demanded and obtained a meeting with state officials to discuss Planafloro. As a result of this meeting, NGOs and the government signed the *Protocolo de Entendimento,* an agreement that, among other things, defined the form and conditions of local groups' participation in Planafloro. The agreement's most important point, from the perspective of the Rondonian groups, was the participation of elected NGOs in the Planafloro Deliberative Council (*Conselho Deliberativo*—CD), the project's highest decision-making body, at least in formal terms.[22] NGOs' seats would be in equal number to those allocated to Planafloro's main implementing agencies. Besides participation in the CD, NGOs would also participate in planning specific initiatives within Planafloro and overseeing budget allocations. As was mentioned above (see note 18), the Rondônia government, in the same document, also committed to the completion of a series of environmental and Amerindian protection measures that had remained incomplete since the end of Polonoroeste.[23] The commitments established in the 1991 agreement were later included in the Planafloro final design, approved in 1992.

It is interesting to notice that the points agreed upon by the Rondônia NGOs and the government in their 1991 meeting reproduced almost exactly the demands stated in the 1990 EDF letter to the World Bank, that is, NGO participation in Planafloro, protection of Amerindian and conservation areas, and compatibility of federal and state policies with the Rondonian socioeconomic and environmental zoning law.[24] This coincidence of demands suggests a strong influence of international groups in the establishment of priorities and definition of strategies by local groups.

By all accounts, the establishment of the Rondônia Forum and its almost instantaneous ability to engage its members in a direct dialogue with the government provoked a general feeling of empowerment among Rondonian civil society groups.[25] To a certain extent, local groups' capacity to organize and speak in a single voice through the Rondônia Forum increased their leverage vis-à-vis the World Bank and, to a lesser extent, vis-à-vis the state and federal governments. The World Bank, in particular, had informally endorsed the creation of the forum and its role in Planafloro, as it would help associate the bank's image with participatory projects. For international and national groups involved in the Rondônia network the existence of a preferential interlocutor in Rondônia greatly facilitated the coordination and legitimacy of activists' strategies. The Rondônia Forum quickly became the primary catalyst for initiatives to guarantee the effective implementation of Planafloro's environmental component.

The forum's strategies remained, through the early 1990s, heavily modeled after those of the MDB campaign. Its members emphasized letter-writing campaigns directed to the World Bank, to Brazilian agencies in charge of implementing Planafloro, and to government representatives at various levels. They also mobilized media in Rondônia and abroad, and to a lesser extent, in the rest of Brazil. Finally, as had happened during the mobilization against Polonoroeste, the forum, with the financial support of its international partners, conducted and commissioned many studies and monitoring reports on the Planafloro implementation.

The strategy of calling upon the World Bank to increase monitoring of Planafloro usually had the effect of temporarily strengthening Rondonian civil society groups vis-à-vis the state government. For instance, one week before the final version of Planafloro was submitted to the World Bank for approval, the Rondônia Forum sent a letter to the bank's board of directors calling attention to the fact that most of the commitments assumed by the Rondonian government in its agreement with the NGOs in June 1991 had not been fulfilled.[26] While the letter did not prevent the approval of Planafloro, it immediately provoked the visit by a World Bank mission to Rondônia to oversee the status of implementation of Planafloro preconditions. The World Bank's interest and its quick answer to the charges of the Rondônia Forum temporarily strengthened the latter's position vis-à-vis the Rondonian government.[27] Bank staff held direct meetings with members of the Rondônia Forum conveying to the government the message that they considered civil society groups as legitimate interlocutors.

The eagerness of World Bank managers to establish a working relationship with the Rondônia Forum is a most interesting indication of cultural and procedural changes within the institution as a result of outside pressures.[28] Some members of the Planafloro World Bank management team—including the project's task manager until 1994—had participated in Polonoroeste. During the five years that separated the two projects, perceptions on the role of NGOs in project implementation seemed to have changed significantly inside the bank. During Polonoroeste, for instance, the World Bank considered the Brazilian Indian agency, FUNAI, as the only representative of Amerindians' interests. In Planafloro, on the contrary, World Bank monitoring missions to Rondônia always engaged in dialogue with forum members, among whom were the leader of the Karitiana Indigenous peoples and representatives from the National Indigenous Union—Rondônia.[29]

The direct engagement of World Bank officials with members of the Rondônia Forum generated, among the Rondonian government and Planafloro's implementing agencies, the perception of a *potential threat* to their tendency of treating the project as a development, rather than an environmental protection initiative. Yet, the *actual threat*, that is, World Bank's effective sanctions against the federal and state governments for failing to adequately implement Planafloro, never materialized. In time, what was once a political asset for the Rondônia Forum (its direct dialogue with the bank) became a liability, as I will discuss below.

One positive aspect of the dialogue between forum members and World Bank officials was that the latter were advised, early on, regarding one of the major threats to Planafloro's environmental component. According the evaluations by forum members, the colonization plans of the Brazilian national institute for colonization and agrarian reform, INCRA, were in direct opposition to Planafloro's objectives.[30] Furthermore, such plans disregarded local and federal legislation that required environmental impact assessments (EIAs) and barred settlement initiatives in ecologically sensitive areas.

Both the Rondônia Forum and the World Bank wrote several letters to the Brazilian government requesting immediate action to guarantee INCRA's compliance with Planafloro's objectives.[31] These letters demanded the cancellation of specific colonization plans, the removal of squatters from areas where conditions were not adequate for human occupation, and the immediate initiation of land redistribution processes in areas defined by zoning regulations as adequate for sustainable agriculture.[32] The letters written by the Rondônia Forum, in particular, contained impressive documentation and evaluations on INCRA's activities in Rondônia. Despite their frequency, the seriousness of the accusations they contained, and their accurate assessment of the agency's illegal behavior, these letters generated formal explanations and vague promises from high-ranking officials in the Rondonian and Brazilian governments, but no concrete action.

Thus, early in 1993, members of the Rondônia Forum realized that the strategies they had borrowed from the MDB campaign (letter writing, press releases, and indirect pressures on the Rondonian government via World Bank monitoring missions) were ineffective. They opted for a change of course and called on the support of their international partners in a lawsuit against the colonization agency.[33]

The carefully researched and documented lawsuit against INCRA convinced the courts of the merits of the environmentalists' charges. The forum's lawyer at the time, who drafted the lawsuit, explained: "The agency has expropriated a huge area, around four hundred square kilometers, for the settlement of landless people. But this is happening in an area designated [by the zoning law] for extractive reserves!"[34] The courts responded to the lawsuit by issuing a provisional order mandating the agency to stop all settlement initiatives in Rondônia.

Following the victory against INCRA in the state courts, the leadership of the Rondônia Forum devised a bold strategy. In June 1994, it wrote a letter to the World Bank requesting the interruption of disbursements for Planafloro.[35] The request was based on evidence that neither the Brazilian nor the Rondonian government was committed to project implementation. Letters from international and Brazilian environmental and human rights NGOs to the bank supported the forum's request, and committed sectors of the national and international media gave full coverage to the issue.[36] Yet, among Rondonian grassroots groups, the request for Planafloro's interruption generated strong opposition.

Seen as a continuum, the forum's lawsuit against the colonization agency and its letter to the World Bank requesting the interruption of Planafloro reveal a lack of synchronicity among the goals and strategies of the Rondônia network's members. This disjunction was even more serious within the Rondônia Forum itself, leading it to experience an "identity crisis" by the end of 1994. The lack of synchronicity is revealed by the network's ambivalence in pursuing either a local or an international strategy to hold INCRA accountable to Planafloro's objectives.

On the one hand, the lawsuit against the colonization agency was a successful attempt by local groups, supported by their international partners, to "localize" the pressures for an adequate implementation of Planafloro. The strategy avoided international arenas and international sources of leverage, and relied entirely on the legal and research skills of local groups. Most important, the primary objective of the lawsuit was to protect the interests of a sector of Planafloro's beneficiary populations (the rubber tappers who would eventually benefit from extractive reserves in the area, and those interested in the state's overall environmental conservation). On the other hand, the letter to the bank requesting Planafloro's interruption took the network's activism back to international arenas. It placed the request of the Rondônia Forum squarely within the context of the MDB campaign, providing ammunition to the campaign's claims that little had changed within the multilateral development banks since the Polonoroeste disaster. In terms of the goals of the strategy, the perception of local groups was that their leaders in the Rondônia Forum were more concerned with making strong statements about the development policies of multilateral organizations than with the imminent difficulties of their constituencies in Rondônia.

When the Rondônia network, after the completion of Polonoroeste, renewed its mobilization efforts in the early 1990s, the need to establish a local membership base in Rondônia capable of implementing local strategies toward environmental protection, was of paramount importance for all network members. The creation of the forum was a partial response to this need and it quickly became a reference point for environmental advocacy initiatives in Rondônia, both for organizations outside and within Rondônia. The forum, however, struggled to find a local identity while still being, to a large extent, a "creature" of international NGOs. The forum's identity crisis would not be resolved until the legitimacy crisis that affected the Rondônia network as a whole in 1994 was itself resolved. The next section discusses this legitimacy crisis in detail, highlighting the difficulties of "localizing" transnational environmental activism. How the Rondônia network eventually overcame such a crisis is the topic of the following chapter.

The Rondônia Network's Legitimacy Crisis of 1994

As soon as the World Bank received the letter from the Rondônia Forum demanding the interruption of Planafloro, it sent a monitoring mission to Brazil. The mission was prepared to accept the forum's demand. According to Schwartzman, of the EDF,

> [T]he Bank was going out of its mind with worry that it would be forced to suspend [Planafloro] and was facing still another massive scandal in Rondônia.... The provisional court order [against INCRA] put the Forum in a massively powerful position vis-à-vis the Bank and the state government. They could have, had they wanted to, suspended that loan and instantly written a ticket for having it re-started.[37]

Yet when the bank mission arrived in Rondônia it encountered a different scenario from what it had expected. Rondonian civil society organizations had replaced the June letter's confrontational tone with a pragmatic willingness to negotiate.

What happened between June and August 1994? The answer lies in the diversity of agendas and expectations among the members of the Rondônia network and particularly among the members of the Rondônia Forum. The request for suspension of Planafloro's disbursements was a bold, yet hastened, strategy that addressed the concerns of the forum's leadership and was supported by its national and international allies. If pursued, this strategy might have turned into an important symbolic victory for those concerned with the environmental sustainability of development policies not only in Rondônia, but in the entire developing world.

Grassroots organizations representing rubber tappers and rural workers in Rondônia, however, were primarily concerned with the expectations of their rank and file members. For them, Planafloro may have been tainted by problems, but it represented these populations' best chance of obtaining concrete benefits (such as rural credit and the demarcation of extractive reserves). The president of the rubber tappers organization, for instance, stated emphatically: "[N]ever, speaking for the OSR, have we been in favor of suspending Planafloro. We were in favor of adjustments in the program but we could not throw it all away. For us it would have been a major loss." The same rationale is voiced by an interviewee close to the rural workers: "[F]or as long as there is money from Planafloro, there will be resources for small farmers. When the Planafloro ends, these resources will end with it."[38]

The Rondônia Forum's leadership thus did not have the necessary political support and internal consensus to pursue the interruption of Planafloro. It opted instead to be responsive to the interests of its members and to negotiate with the government and the World Bank in August 1994. These negotiations, however, resulted in yet another aide-mémoir that simply redefined schedules and restated commitments from different agencies within Planafloro.

In retrospect, the decision to negotiate was a "strategic error."[39] In the post-negotiations phase, the forum and its member organizations faced a series of challenges. First, they needed to strike a balance between their contradictory roles in co-implementing and independently monitoring Planafloro. Second, they had to be responsive to new project implementation commitments that they assumed during the August 1994 negotiations, which they soon discovered to be beyond the limits of their technical capacity.[40] Third, they needed to regain the trust of national and international allies, who had supported the request for Planafloro's interruption, only to see it withdrawn a few months later. These challenges clearly illustrate the argument that participation in transnational advocacy networks, while generating short-term political empowerment for local groups, may bring further responsibilities that these groups may not always be prepared to shoulder.

With the request for Planafloro's interruption, the Rondônia Forum made it evident to the government that the forum alone could stop funding for the project. Yet, once the forum showed its political force, it had to occupy the political space it had conquered. Its members chose to do so by negotiating with government and World Bank officials. The political weight and legitimacy that the forum and its member organizations had acquired created conditions that deepened their participation in Planafloro implementation. Thus, they assumed further responsibilities within the project. Some of these responsibilities were mere reinstatements of previous ones, such as participation in the Planafloro deliberative council and in different instances of budget allocations for the project. Yet, there were some new responsibilities as well, such as co-participation with local agencies in monitoring invasions of conservation units and Amerindian reserves. During 1994, it became evident that the forum and its affiliated organizations could not meet these challenges.

The political leverage that members of the Rondônia Forum gained in local environmental and development policymaking arenas as a result of their participation in the Rondônia network was not accompanied by an equivalent increase in their technical capacity. Local groups' technical capacity was extremely constrained by limited financial resources. To a certain extent, this factor helps to explain the Rondônia Forum's internal cleavages between advocacy NGOs organizations and grassroots groups. Most NGOs in Rondônia at the time depended entirely on funding from international organizations. The Rondônia Forum itself, even today, has no independent source of funding. Throughout its history it has been supported by grants from Oxfam, World Wildlife Fund for Nature, the Damien Foundation, and the Rainforest Action Network.[41] The Indigenous Peoples' Missionary Council also had funding from Oxfam, to cite just one other example.[42] It is not surprising, thus, that leadership in support and advocacy NGOs was sensitive to the agendas of their international partners, and more aware of international environmental advocacy initiatives than were many grassroots groups in Rondônia.

Many grassroots groups, on the contrary, in part thanks to Brazilian legislation on labor organization, have a guaranteed source of funding for their operations. Organizations such as local chapters of rural workers' unions and the Federation of Agricultural Workers of Rondônia (FETAGRO) receive compulsory financial support from their affiliates. Still, the financial resources promised by Planafloro to its beneficiary populations (Amerindians, rubber tappers, small farmers, and riverside communities) had an enormous appeal. It explains the pressures that these populations placed on their representatives and on the Rondônia Forum to negotiate better conditions for project implementation, rather than terminating the project.

As a consequence of their low budgets, Rondônia NGOs and grassroots groups had great difficulty in attracting legal experts, financial consultants, and economists. Without such experts, civil society participation in budget allocation committees for Planafloro and in monitoring the state's environmental legislation was practically meaningless. Representatives of civil society organizations whom I interviewed in Rondônia in late 1994 systematically mentioned their extreme need for legal and economic experts. At that time, the Rondônia Forum had hired a lawyer and its member organizations also counted on the occasional support of a Brasília-based NGO that specialized in environmental and Amerindian rights.[43] The lawsuit against INCRA only became possible because of these resources. They were, however, still too limited to allow local groups to systematically pursue legal avenues to uphold the rights of their constituencies.

While the technical shortcomings of Rondonian civil society groups limited their capacity to influence Planafloro's implementation, it was the project's own structural features that entirely compromised the process.[44] But because many Rondonian NGOs and grassroots groups were, on paper, co-responsible with state and federal agencies for the project's implementation, public disenchantment with the project eventually affected them. It is ironic that the very success of the Rondônia network in including participatory mechanisms in the Planafloro design later contributed to undermine the credibility of local civil society organizations vis-à-vis their rank-and-file. By the end of 1994, the general feeling among Rondônia NGOs and their constituencies, reflected in the statements below, was that nothing had been accomplished under Planafloro.

> The Rondonian NGOs, today, are undergoing an identity crisis, a huge operational crisis, the Forum is in a bad shape. It is essentially an internal crisis. It may be able to maintain an external image, but internally, it is disastrous![45]

> We have arrived at December 1994 without the demarcation of a single extractive reserve or indigenous area. We have wasted energy, money, a lot indeed, yet the movement has accomplished nothing. Worse still, it has lost credibility, it is empty now . . .[46]

> Planafloro's delay in establishing the extractive reserves is causing the rubber tappers movement to be discredited.[47]

> Planafloro has promised the Indians a lot, but it hasn't delivered anything! Up until now, Planafloro has served only as a marketing tool. It promised to place forty doctors in indigenous areas and a high number of nurses, but it all remains as it was, and health problems have aggravated in indigenous areas.[48]

Pragmatism ran deep within some groups. By November 1994, the Organization of Rondonian Rubber Tappers, for instance, had started independent negotiations with the colonization agency in a desperate effort to guarantee Planafloro's resources for the establishment of extractive reserves. This move was justified if the OSR were to preserve its institutional credibility among rubber tappers. During the Planafloro negotiations, the OSR leadership went to rubber tappers' communities to discuss their needs and to announce the creation of reserves under Planafloro. It at once raised the expectations of its rank and file and encouraged the perception that it was itself associated with the project. Planafloro's lack of results almost two years into its implementation harmed OSR's relations with its constituency. The latter became skeptical about the project's promises, and, as a consequence, about those voiced by the OSR.[49]

The low levels of organization of many local groups, particularly Amerindians, also negatively affected the credibility of civil society organizations in

Rondônia. An interesting example was the tension created by logging operations in Amerindian reserves. The Rondônia Forum systematically pressured authorities to organize surveillance missions to Amerindian areas to prevent illegal logging. Naturally, Amerindian organizations, such as UNI and the Karitiana Indigenous Peoples' Organization, as well as organizations supporting indigenous rights, such as the CIMI, were among the leading groups involved in these initiatives. Staff of these organizations even participated in monitoring missions on many occasions. Yet most Amerindian groups in Rondônia had (and continue to have) agreements with logging companies that allow operations in their lands. It was not unusual for some Amerindian groups to take a proactive position against Planafloro's monitoring missions.[50]

Finally, not all local civil society organizations were prepared to evaluate and provide insights into the highly complex nature of Planafloro's environmental issues. The Rondônia Forum had difficulty democratizing debates within its own ranks. Some groups resented the "academic level" of the discussions that occurred in meetings at the forum's headquarters, and were not fully able to participate in technical debates with bank officials.[51] What one experienced by the end of 1994 in Rondônia was a profound sense of desolation due to unfulfilled promises. Not only had Planafloro failed to deliver the benefits it once anticipated, but one of the most daring experiments in civil society mobilization in the Brazilian Amazon region, the Rondônia Forum, was at the brink of collapse. In this context, how does one evaluate the impact of the Rondônia network in the early 1990s?

Taking Stock of the Rondônia Network in the Early 1990s

The greatest achievement of the Rondônia network in the early 1990s was the establishment of a local membership base fully integrated into the network's initiatives and constitutive principles. This is not to say that relations among international, national, and local members of the network were always smooth or that cleavages of different natures did not exist among them. Yet, in comparison to the formative years of the network, the early 1990s witnessed a growing influence of local groups over other network members. One of the consequences of this increased influence was the network's commitment to broader (and long-term) goals of environmentally sustainable development, with less focus on the immediate consequences of environmental degradation.

Coherent with its longer-term approach to environmental protection, the Rondônia network invested valuable resources toward the establishment of permanent mechanisms for local environmental monitoring. The creation of the Rondônia Forum was in part a response to the frustrations that members of the network had experienced when trying to monitor environmental conditions in Rondônia from a distance during the Polonoroeste project. Yet the forum was not entirely a "creature" of organizations located outside Rondônia. It was also one of the consequences of an accelerated awakening of local civil society groups to the need to assert the environmental and citizenship rights of their constituencies.[52]

Given the challenges of the context, it took the Rondônia Forum some time to establish its institutional identity. Perceptions of the role, political weight, and legitimacy of the organization in its first two years of existence varied widely. It is true that some analysts and network members perceived the Rondônia Forum mostly as a reference for external actors, namely, international environmental NGOs and the World Bank. Their assessment is justified by the fact that, at its inception, the forum counted on limited participation of local grassroots groups. Most significantly, it operated within a preestablished framework (that provided by the campaign against the Multilateral Development Banks). But even these observers noted that, for certain Rondonian groups, the forum was a valid resource in their struggle for citizenship.[53]

For others, the forum was an arena for the aggregation of interests and collective action strategies for Rondonian groups sharing a common vision of Rondônia's development and the role of Planafloro in it. As such, the forum was politically strong in its negotiations with World Bank officials, who perceived the project as an environmental initiative. The forum, however, remained politically weak vis-à-vis the Rondonian government, which prioritized the developmental aspects of the project and saw no role for civil society organizations in it.[54] Finally, there were those who saw the Rondônia Forum as a valid, yet not sufficient, interlocutor of the expectations of the Rondonian civil society toward Planafloro.[55] As such, it was a political instrument of support and advocacy NGOs, but not an arena for the discussion of the interests of Rondonian grassroots groups and social movements.

The opportunities that the Rondônia network generated for local groups at the onset of Planafloro temporarily overshadowed the challenges of participating in and monitoring the project. From the start, these were contradictory roles. If local groups were to be co-participants in implementing Planafloro, could they also independently monitor it? In fact, because the implementation of Planafloro's environmental component was all but paralyzed during 1993 and 1994, local civil society groups adopted the default role of monitoring the project. Yet that strategy was not sufficient to eliminate perceptions of the formal links between members of the Rondonian network and the Planafloro project that prevailed among certain sectors of Rondônia's population. As the project became discredited, so did many local civil society groups.

Finally, the process of "localizing" the Rondônia network was affected by a certain degree of unintentional ambivalence from the part of its leading members. On the one hand, the network as a whole praised the initiative of local members in devising local strategies to guarantee Rondônia's environmental integrity. Such was the case of the 1994 lawsuit against INCRA. Yet, rather than channeling further resources to encourage similar initiatives, the network catalyst organizations opted for supporting a return of activism to international arenas, as was the case of the campaign for the interruption of World Bank disbursements for Planafloro.

In the following chapter I discuss the ways by which many of the challenges of localizing transnational environmental activism were resolved within the Rondônia network. Yet, in a less than perfect world, many challenges still remained and others emerged. At the turn of the millennium the Rondônia network achieved political maturity and provided lessons that may contribute to the promotion of environmentally sustainable development in Amazonia and beyond.

5

Listening to the Grassroots— The Rondônia Network and Local Politics

The previous chapters focused on the internal politics of the Rondônia network. Without abandoning that focus, the present chapter expands the scope of the analysis to evaluate the consequences of transnational activism for local politics. Over time, the process of "localizing" the Rondônia network has politically empowered sectors of the local civil society. Empowerment has occurred within a political context that remains characterized by the weakness of democratic institutions and by the control by economic elites of the state's resources. In such a context, the concept of empowerment must be qualified. In the case of Rondonian civil society organizations, political empowerment means, among other things, establishing a position of legitimate interlocutors in the eyes of the state, enhancing organizations' ability to obtain and disseminate information on public policy, securing—by pressure and dialogue—the implementation of emergency measures to safeguard the environment and local Amerindian populations, blocking in local courts attempts by the government to undermine environmental protection gains, and, most important, obtaining direct access to financial resources originally under the complete control of the state's bureaucracy. In addition, the empowerment of local civil society groups includes increasing their understanding of both policymaking processes in Rondônia and the links between development and environmental sustainability.

The story of the Rondônia network's activism at the turn of the millennium illustrates the different ways in which participation in transnational environmental advocacy networks may contribute to the political empowerment of local groups. It also illustrates how this process has created new challenges for local groups and how the failure to address these challenges has sometimes led to significant setbacks, both for the objectives of the network as a whole and for local groups in particular. Finally, the story of the Rondônia network raises questions about the relations between transnational environmental activism and local political struggles for citizenship rights, such as whether there are clear boundaries between the two processes, and to what extent they interface. While this chapter does not offer complete answers to these questions, it highlights their theoretical importance for studies on transnational environmental activism.

This chapter addresses the two main processes by which the members of the Rondônia network overcame its 1994 legitimacy crisis. It also evaluates the impact of such processes, both within the Rondônia network and outside it, on the larger context of the state's politics. The first process was the effort to bring the Planafloro project to the attention of the World Bank Inspection Panel. The panel, created in 1994, is composed of three independent experts, who "investigate complaints submitted by people directly affected by Bank projects regarding violations of World Bank policies, procedures, and loan agreements."[1] The second process was the formulation of an alternative approach to environmentally sustainable development in Rondônia and to the role of Planafloro in it. Through this alternative approach, environmental sustainability was not a goal in itself, but an integral part of the struggle for improvement of the quality of life, in environmental, socioeconomic, and political terms of sectors of the Rondonian population. This new approach became predominant within the Rondônia network in the late 1990s, replacing the earlier and narrower emphasis on conservation units and protection of Amerindian lands.

The Rondônia Network Takes Planafloro to the Inspection Panel

In June 17, 1995, the Rondônia network presented to the inspection panel a request for the investigation of the Planafloro project. The main charge was that the bank had failed to comply with its monitoring policies. Well into its third year of implementation, Planafloro still had not accomplished most of its environmental goals. The World Bank, however, continued disbursing funds for the project, as originally scheduled.

The Planafloro inspection panel claim surprised both the World Bank and the Rondonian government.[2] Practically all of the most significant groups representing the Planafloro beneficiaries (the Organization of Rondonian Rubber-Tappers, the Federation of Agricultural Workers of Rondônia [FETA-

GRO], the Rondonian chapter of the MST, and the Coordination of the Union of Indigenous Peoples and Nations of Rondônia [CUNPIR]), and many important civil society organizations concerned with environmental and human rights in Rondônia (Ecological Action for the Guaporé Valley [Ecopore] and the Institute for Pre-History, Anthropology, and Ecology [IPHAE]) signed the claim. Among its several consequences, probably the most significant was its impact on Rondonian politics. The claim represented a watershed in state–civil society relations. It brought to a climax the long-standing and increasingly fierce contentious relationship between the development model sponsored by the state and that envisioned by sectors of the local civil society.

How was it that the panel strategy obtained such a level of consensus among Rondonian organizations that, less than a year before, had been divided by profound cleavages? Even more surprisingly, how was it that the strategy succeeded in re-engaging international environmental groups in efforts to promote environmentally sustainable development in Rondônia? The reader should recall the frustration of some international NGOs with the negotiation option that their local partners adopted following the 1994 request for Planafloro's interruption.

The answer to both of these questions lies in the catalyst role of the inspection panel strategy. Local groups rallied behind the claim not only for its novelty among several "worn out" strategies, but also because its ultimate goal—to improve Planafloro's implementation—addressed their immediate interests. International groups embraced the strategy both for its potential significance for the environment in Rondônia (the possibility of actually making Planafloro fulfill its environmental mission) and its impact on larger policies of multilateral development institutions (the opportunity to "test" the validity of the panel as an instrument for World Bank accountability vis-à-vis project beneficiaries).

To understand how the Rondônia network regained its internal cohesion and, consequently, bolstered its legitimacy within both local and international contexts, one must return to the months following the 1994 negotiations between the Rondônia Forum, the state, and the World Bank. The 1994 crisis triggered two parallel processes among Rondonian civil society groups. One related particularly to the Rondônia Forum, whose staff was forced by the crisis to reassess the organization's identity and mission. The other related to the majority of Rondonian grassroots groups, who eventually came to realize that negotiations and cooperation with the government on Planafloro had led them to a dead end.

The Rondônia Forum's reevaluation of its role among local civil society groups and within the Rondônia network resulted in an increased degree of accountability to Rondonian grassroots organizations in general, and to those representing Planafloro's beneficiaries in particular. In fact, the forum's increased accountability to Rondonian grassroots organizations was a natural development of the evolution of its membership base. Table 5.1 shows that between 1991, when the forum was created, and 1995, its membership changed considerably. While in 1991, "brokerage" organizations, that is, advocacy and support NGOs (usually with close contacts with and funded by international groups), constituted the large majority of the forum's affiliated organizations (nine), and only two entities represented grassroots groups, in 1995, advocacy organizations constituted less than half of the forum's members, while the number of grassroots organizations had increased by 100 percent. This tendency toward attracting grassroots organization has intensified, and in 2000 these organizations composed two-thirds of the forum's membership base.

As it became evident that the Rondonian government would continue to ignore its commitments to Planafloro's environmental provisions in the period

TABLE 5.1
Membership in the Rondônia Forum by Year

Year	Advocacy/Support Organizations*	Grassroots Organizations**	Total Membership
1991	9	2	11
1992	11	6	17
1995	12	13	25
2000	11	21	32

*Defined as organizations that provide material, technical, and political support to grassroots groups but are not identified by the "rank and file" of such groups as co-participants in their way of life and struggles, or as their representatives. They are usually professionally organized, i.e., have formal headquarters, communication resources, and paid staff. Examples in Rondônia are Ecopore and the Environmental Protection of the Cacoal region (PACA).

**Defined as organizations that represent a segment of the population that shares similar interests and/or life conditions. Grassroots organizations may or may not have formal headquarters and paid staff. Examples in Rondônia are the OSR and various cooperatives of small farmers.

SOURCES: *Protocolo de Entendimento ONGs—Governo de Rondônia*, June 20, 1991, *Carta Aberta* (from the Rondônia Forum to the World Bank Executive Directors), March 12, 1992; Letter to Fernando Collor de Mello, President of Brazil, May 29, 1992, Request for Inspection, submitted to the World Bank Inspection Panel on the Planafloro, July 25, 1995, and *Forum das ONGs de Rondônia* (institutional brochure), May 2000.

following the 1994 negotiations, Rondonian grassroots groups reevaluated their expectations. A meeting organized by the Rondônia Forum in December 1994 triggered this process. The forum's executive secretariat at that time defined this reevaluation process as the "end of the romantic phase" for civil society groups, who ceased to have any expectations about the project.[3]

It was in the midst of this search by local groups to redefine their role in Planafloro that, in early 1995, international NGOs, such as Friends of the Earth and Oxfam, suggested the inspection panel claim strategy. Their suggestion fell on fertile ground.[4] Though the research for the Planafloro claim was financed, initiated, and essentially completed by staff and consultants in international environmental NGOs (Friends of the Earth, Dutch Organization for International Development Cooperation [NOVIB], and the Center for International Environmental Law [CIEL]), the decision to request the inspection panel's investigation was made together by both local and international groups.[5]

The inspection panel claim was, possibly, the most powerful and successful strategy implemented by the Rondônia network to date, for several reasons. First, the panel had just been established and its effectiveness was under close scrutiny on the part of the international environmental and human rights movements. Addressing the panel meant the possibility of refocusing the attention of the national and international media back to Rondônia.[6] Second, the claim represented a new strategy available to the Rondônia network, at a time when other strategies (letters, negotiations, follow-up agreements) had become ineffective. Third, the mobilization, research, and consultation processes that preceded the filing of the claim reunified groups within and outside Rondônia. This new degree of internal cohesion contributed to increase the political space and political power of Rondonian civil society groups vis-à-vis the local government.[7] Several sources acknowledged the political cohesion of Rondonian civil society organizations during the inspection panel process. An outside observer explains: "the forum was experiencing a crisis. The inspection panel claim reunified groups inside and outside Rondônia."[8] In Rondônia, representatives of grassroots organizations openly declared their support for the requested inspection, regardless of its consequences. The rubber tappers' leader affirmed: "I was favorable to the inspection panel claim from the beginning ... even if it meant interrupting the project. . . ." And the president of FETAGRO recalled: "During the inspection panel claim, we had a common position with these groups (environmentalists, indigenous peoples, and rubber tappers)."[9]

The Planafloro inspection panel claim essentially denounced the World Bank's failure to adequately monitor project implementation. For instance, the Planafloro loan agreement required the implementation of an institutional reform program to make federal and state policies compatible with socioeconomic and ecological zoning and with principles of sustainable resource management. The Brazilian and Rondonian governments had failed to implement such reforms, and the World Bank had neglected to enforce compliance.[10] The incapacity and unwillingness of federal and state governments to harmonize development policies with Planafloro's objectives compromised the project's performance in four main areas: land tenure, the creation of conservation units, the implementation of environmental protection measures, and support to indigenous communities.[11] A fundamental aspect of the Planafloro claim, and the key to understanding why it became such a catalyst in resolving the Rondônia Forum's legitimacy crisis, was its main goal. Rather than expecting that a panel's investigation would lead to Planafloro's interruption, the Rondonian claimants wanted the investigation to "contribute to the solution of the ongoing problems of execution of Planafloro . . ."[12]

The Impact of the Planafloro Claim

The Planafloro claim, as a case study, reveals the impact of transnational environmental activism at both international and local levels. Internationally, it initially challenged the credibility and effectiveness of the panel as an accountability mechanism for international finance institutions. In the longer term, it has triggered processes of redefinition of eligibility criteria and reevaluation of this mechanism. Domestically, the debate generated by the Planafloro claim created opportunities that ultimately increased the political leverage of local civil society groups vis-à-vis the Rondonian government. These impacts can be fully appreciated by following the reactions of the major actors affected by the claim, namely the World Bank and the Brazilian and Rondonian governments.

Chronologically, the first set of responses to the claim came from the World Bank Planafloro management team. It presented to the board of directors a document arguing for the claim's dismissal on the basis of legal and technical irregularities. Management's response, however, did not challenge the essence and merit of the charges made by the members of the Rondônia network. One of management's key arguments was that claimants had failed to fulfill the panel's resolution requiring proof that the bank's actions had caused "material adverse harm." As the argument went, while Planafloro had not produced *potential/planned* benefits to target populations, it had not caused them *actual* "material adverse harm."[13] In subsequent responses to management's arguments, Rondonian organizations rejected their narrowness, and described the links between social and economic processes encouraged by Planafloro and the deterioration of the quality of life of beneficiary populations. It is interesting to notice that, while denying the charges contained in the claim, the management team intensified its oversight of

Planafloro, and pressured both the Brazilian and the Rondonian governments to implement key environmental measures. At least one analyst has described these actions, and others that unfolded on the same path, such as management's Action Plan of 1996, as "schizophrenic." He asks: "Why is management deploying efforts to fix problems whose existence it denies?"[14]

The Planafloro claim challenged, in an unprecedented way, the World Bank executive board. The board usually reaches decisions by consensus. The Planafloro claim, however, created several divisions among directors. On the one hand, as David Hunter, of CIEL, summarizes, there was the dilemma created by the "difficult choice of embarrassing Brazil, a major client/borrower [and a voting part in the process], or rejecting a strong claim, thus crippling the Panel."[15] On the other hand, the claim had created a North-South split that prevailed for several board meetings at which Planafloro was discussed. Directors representing countries in the North, with the exception of those from France and the United Kingdom, supported the Planafloro investigation. They were influenced by both the merits of the claim and the lobbying efforts of Rondonian leaders and their international allies, who met with several directors in Washington, D.C., during the October 1995 annual meeting of the World Bank and International Monetary Fund.[16] Opposing the claim were most of the executive directors who represented Southern countries. They were led by a coalition of directors from Brazil, India, and China, with the then Brazilian director, Marcus Caramuru de Paiva, at the center of the mobilization. The prospect of an investigation by an external body of experts was perceived by Brazil as an attack on its national sovereignty. Paiva was determined to use Brazil's diplomatic weight to avoid an investigation of Planafloro, and clearly conveyed this message to the Rondonian leaders.[17] Eventually, the divisions within the board of directors were resolved by approving management's action plan in lieu of an investigation. In other words, the Planafloro Inspection Panel claim had been rejected.

As I stated at the beginning of this section, among the impacts of the Planafloro claim were the dilemmas and precedents that it created and that eventually influenced changes within the inspection panel itself. The first precedent was the acceptance of management's action plan as a substitute for an investigation. The plan simply proposed an increased oversight of Planafloro. Since the Planafloro claim was rejected, action plans have replaced full investigations in four out of six requests for inspection.[18] As a consequence, panel claimants have received, as a response to their concerns, a remedy created by the very actors who allegedly generated such concerns, rather than an independent assessment of problems. The challenges generated by the Planafloro claim also strengthened efforts by a group of executive directors, led by Brazil's, to restrict the panel's mandate. In the context of the 1999 review of the inspection panel, Brazil sponsored a proposal to dismiss claims whose charges related to a project's failure to improve beneficiaries' conditions. According to the proposal, the panel should restrict its investigations to projects that have caused actual material harm.[19]

The reaction of the Rondonian government to the inspection panel claim was immersed in contradictions. Such contradictions were the result of pressures that pulled the newly elected administration of Governor Walter Raupp in different directions. The first set of pressures that Raupp faced came from the World Bank, by means of increased levels of monitoring, including visits to the area by high-ranking bank officials and even by executive directors.[20] The second set of pressures came from the Brazilian federal government, which wanted to avoid, at all costs, further international scrutiny of development policies and practices in Rondônia. The pressures from the federal government were taken seriously by the Raupp administration, since the cancellation of Planafloro would lead to the paralysis of all new loan negotiations for the state.[21] The third set of pressures endured by the Raupp administration came from grassroots groups who had signed the Planafloro claim, such as MST and FETAGRO. These groups had political links with the *Partido dos Trabalhadores* (The Worker's Party), a party that had become part of the ruling coalition by endorsing Raupp's candidacy in the state's runoff elections. Finally, the Rondonian government remained constrained by its links to the state's economic and political elites.

In light of these pressures, it is not surprising that the Rondonian state and federal agencies rushed to comply with Planafloro's environmental conditions. On July 28, 1995, federal and state land agencies signed an agreement transferring federal lands to the state of Rondônia. The agreement had been a precondition to the Planafloro loan and was a necessary step in the process of creating several conservation units and extractive reserves. It was also an important measure to reduce the free reign of the National Institute for Colonization and Agrarian Reform in Rondônia. In the month following the land agreement, the Rondonian government created fifteen extractive reserves. Before the end of the year, the geographic limits of most indigenous areas in Rondônia were physically determined (a process known in Portuguese as *reaviventação*). In the course of 1996, a "notable progress" was observed in the management of conservation units and in the establishment of social and economic infrastructure in extractive reserves.[22]

The speed with which some of these actions were implemented raises questions on the role of the inspection panel claim in this process. The prevailing explanation is that the inspection panel claim and the consequent lobbying by the World Bank of the Brazilian and Rondonian governments were instrumental for the adoption of such important measures (land agreement, demarcation of extractive reserves and indigenous lands).[23] A concurrent explanation emerged from my interviews. The fact that these measures became effective shortly after the filing of the claim indicates that the decision to implement them was being considered even before the request.[24] Thus, the implementation of environmental measures by the Rondonian government was a response to a timely combination of factors: the years of advocacy by local and international groups, the pressures from the World Bank, which had grown increasingly dissatisfied with project implementation, and the catalyzing impact of the inspection panel claim.

After surrendering to the pressures to implement Planafloro's environmental measures, the Rondonian state started to look for loopholes in order to "pacify" its traditional clients, that is, the state's economic elites, particularly logging interests. During 1996, the Raupp administration issued a series of regulations that directly undermined Planafloro's environmental accomplishments. Rondonian civil society organizations, however, were able to block all but one of such attempts, further evidence of their political strength in the wake of the inspection panel claim. For instance, in February 1996, Raupp issued Decree 7,341, allowing logging in areas designated by the state's zoning law for extractive reserves. In less than thirty days, however, the government was forced to reverse its decision, having been defeated in a civil action suit initiated by the Rondônia Forum. In another episode, the governor approved Law 152 in June 1996. The law regulated the continuation of the zoning process but it contained serious irregularities. The Rondônia Forum immediately alerted the World Bank, whose lawyers were able to intervene and correct procedures. Before the end of 1996, the governor issued yet another decree potentially harmful to environmental interests (Decree 7,634/96). The measure added FIERO, the Rondonian business association *(Federação das Indústrias do Estado de Rondônia)*, to a committee in charge of overseeing initiatives to prevent invasions and exploitation of natural resources in conservation units. The Rondonian groups opposed the decision on the grounds that the participation of FIERO, given its links to the logging industry, constituted a conflict of interest. FIERO was eventually denied participation and representatives of rubber tappers and indigenous peoples were brought in. Finally, members of the government's coalition sent to the legislative assembly law projects to reduce the areas of two state parks and one extractive reserve. While these law projects have yet to be approved, they have set a serious precedent for further reduction of conservation units.

The escalation of tensions between the Rondonian civil society organizations and the state reinforced the decision of the Planafloro World Bank management team to facilitate or mediate a dialogue between the parties.[25] This goal was pursued through several initiatives that culminated in the June 1996 Planafloro Evaluation Seminar (which is discussed in the following section). Important steps leading to the seminar included the World Bank–sponsored workshop on Planafloro's participatory management in March 1996 and the bank's commissioning of a long overdue evaluation of the project, conducted by independent consultants in April 1996. In fact, John Garrison, the World Bank liaison with civil society in Brazil, stated that these initiatives were a direct result of the inspection panel claim and "a function of the competence that the [Rondonian] civil society demonstrated in questioning the way Planafloro was being implemented. . . . Through the inspection panel claim, the World Bank changed its perception of Planafloro. The bank realized it had to change strategies; . . . to reach a 'win-win' situation there had to be collaboration [among government and civil society]."[26]

The workshop on Planafloro's participatory management was important as a preliminary forum in which civil society groups "bashed" the World Bank and voiced their grievances about the obstacles to participation. Garrison, again, explains: "this was important since, when we got to the evaluation seminar, [this part of the process] had already occurred." Because the confrontational atmosphere between the World Bank, the Rondonian government, and local civil society groups had been somewhat ameliorated by previous initiatives, the dialogue during the Planafloro evaluation seminar could focus on the technical aspects of project implementation. Civil society groups had the upper hand in this process for two reasons. First, the seminar's working document was the final report of the March 1996 evaluation. The evaluation had reiterated most of the charges contained in the inspection panel claim and thus further legitimated the position of Rondonian civil society organizations. Second, the technical and political limitations of the Rondonian government became quickly apparent at the onset of the seminar. Garrison recalls: "The forum was well prepared. NGOs had better proposals, better formulated and with more details than those from the government. There were speeches from representatives of several sectors of the population, and when it was the government's turn, the secretary of planning did not say anything. The government lacked the capacity to make a competent defense against the NGOs. So, it decided to negotiate."[27]

The cohesion among the members of the Rondônia network immediately preceding, and throughout the duration of the Planafloro inspection panel claim, was instrumental in building up their power vis-à-vis the World Bank and the Brazilian and Rondonian governments. The innovative nature of the strategy and the visibility that a newly created international mechanism such as the panel provided, were also factors in this process. In the end, despite the rejection of the request for inspection, the achievements of the Rondônia network were undeniable. Many measures for the protection of conservation areas and for the control of processes of environmental degradation were finally adopted. While much remains to be done, these measures provide the foundation for future efforts toward environmental preservation in Rondônia. Arguably, even more important than the implementation of Planafloro's environmental conditions was the level of political empowerment that Rondonian civil society groups derived from the inspection panel experience. Not only did they speak with a single voice throughout the process, they also assumed leadership of the initiative (the claim was, essentially, presented by the Rondônia Forum), lobbied without intermediaries in international arenas (such as during the 1995 World Bank/IMF annual meeting), and kept the state government under close scrutiny, preventing it from limiting the scope of the recently implemented environmental measures. Yet the most striking evidence of local groups' political empowerment, following the resolution of the Rondônia network's legitimacy crisis and its experience with the inspection panel, were the negotiations that occurred in the context of the 1996 Panafloro evaluation seminar and its outcomes.

Evaluating and Restructuring Planafloro: A New Meaning for Environmentally Sustainable Development?

In chapter 4 I argued that the activism of the Rondônia network in the early 1990s contributed to consolidate Planafloro's conservationist approach to environmentally sustainable development. Both activists and World Bank and Brazilian negotiators prioritized the immediate protection of conservation units and Amerindian lands, along with the eradication of policies that encouraged deforestation. Planafloro, as originally designed, had not contained provisions to fully reconcile environmental protection measures with the expectations of improvements in the quality of life of project beneficiaries in environmental, economic, and political terms.

As Rondonian groups gained influence within the Rondônia network, issues such as the economic costs and benefits of environmental protection for local populations came to the forefront of the network's agenda. Key constituencies in this process were small farmers and rural workers. These groups had originally been supportive of Planafloro due to its promises of incentives for agroforestry and livestock activities, and provisions for land redistribution. In fact, small farmers and rural workers saw the environmental component of the project as a threat to their potential gains and did not support it. Eventually, however, they changed their approach, as it became increasingly evident that the state was using Planafloro's resources, to a large extent, to favor dominant economic interests in the region, namely, large landowners and logging groups. The Rondônia Forum played an important role in the process of attracting the cooperation of groups such as the landless movement, the federation of agricultural workers of Rondônia, and smaller agricultural cooperatives. Because leadership in these groups participated in many discussions about Planafloro, they had the opportunity to exchange experiences with other groups about the links between environmental degradation and loss of land productivity, and the potential of environmentally sound practices to improve the health and overall material condition of rural communities. This process of environmental education of leaders slowly "trickled down" to their rank-and-file.

As local civil society groups discussed their expectations for Planafloro and established bridges among their once opposing interests, they began to formulate, maybe unintentionally, an alternative approach to Rondônia's environmentally sustainable development (to that offered by Planafloro). This alternative approach attempted to reconcile measures of environmental preservation with complementary activities, for the improvement of the material conditions of Rondonian communities. Civil society groups and selected state agencies had identified a possible mechanism toward this goal within Planafloro, even before the filing of the request for inspection. A miniscule subcomponent of the project, the community initiative program, had been relatively successful in sponsoring environmentally sound initiatives, while actually transferring Planafloro's resources directly to the hands of its primary beneficiaries.[28]

The intensity with which local groups in Rondônia came to embrace the concept of a community initiative program is emblematic of the two priorities that defined the efforts of the Rondônia network in the late 1990s: to protect the state's environment by promoting environmentally sound but economically viable alternatives, and to obtain for Planafloro's beneficiaries their due share of project funds. The fact that these priorities became predominant within the Rondônia network and thus determined the course of the processes of evaluating and restructuring the Planafloro project in 1996 is indicative of the growing influence of local groups both within and outside the network.

Following the World Bank–sponsored workshop on Planafloro participatory management and the release of the independent evaluation report on the project, the World Bank and the Rondônia Forum organized the Planafloro Evaluation Seminar, in June 1996. Rondonian civil society groups arrived at the seminar with their homework done. During the time that elapsed between their recognition of the limits of the inspection panel strategy and the beginning of the seminar, local groups had capitalized on the political visibility they had regained in 1995. They proactively engaged Brazilian environmental experts and advocacy think tanks in proposing viable mechanisms to enhance Planafloro's contribution to the state's environmentally sustainable development. They were thus fully prepared to present a proposal for Planafloro restructuring that the World Bank supported and that the Rondonian government was not able to reject.

The Rondonian government's lack of alternatives for Planafloro's restructuring became evident at the onset of the seminar. Yet technical incapacity was not the only aspect of the state's participation in the evaluation seminar. Representing the government was the Rondonian secretary of planning, Emerson Teixeira, who had once been affiliated to the Worker's Party and was sensitive to the demands of local civil society groups.[29] At his suggestion, the evaluation seminar was interrupted to allow negotiations between the government and civil society groups to take place. World Bank officials mediated this process, while consultants from national and international organizations advised the Rondonian groups. Negotiations focused on two related demands from civil society. The first was that Planafloro funds be streamlined and several governmental bureaucracies excluded from it. The second was that a portion of the reallocated funds be directed to expand the community initiatives program. At the conclusion of negotiations, the Rondonian civil society organizations had achieved their goals. Many state bureaucracies had been excluded, and one-third of the Planafloro remaining funds (approximately $20 million) were directed to a community initiative fund, the Program for the Support of Community Initiatives (*Programa de Apoio a Iniciativas Comunitárias*, or PAIC). The Rondonian government, however, maintained control over the Planafloro in terms of overall administration and responsibility for its remaining components, in this case, environmental conservation and transportation infrastructure.[30]

Following the evaluation seminar, the main challenge for the Rondônia network, and in particular for its local membership base, was to redesign the expanded community initiative program. The Rondonian groups embraced this task in part as an opportunity to restructure Planafloro in terms of their own approach to environmentally sustainable development. As they envisioned it, the community initiative program was to be an opportunity for the project's beneficiaries themselves to define their development priorities, aiming at the improvement of their quality of life in economic, political, and environmental terms.[31]

The Community Initiatives Program and Environmentally Sustainable Development in Rondônia

The *Manual of Operation* for PAIC describes its general objectives as "to fund and implement projects conceived from the perspective of the needs, initiatives, and priorities as defined by the beneficiary communities themselves, aiming at the improvement of their quality of life, employment opportunities, income, technology and production, strengthening of their citizenship, social organization, participation, self-management, and environmental conservation, in line with the Planafloro's objectives" (p. 5, author's translation).

As the reader can see, environmental issues per se are not first among the program's priorities; the improvement of beneficiaries' quality of life is. Yet a closer analysis of the program indicates that environmental conservation is recognized as an intrinsic part of a satisfactory quality of life. The program is conceived as a vehicle for the consolidation of Planafloro's environmental goals, and a tool in the process of raising the environmental consciousness of its beneficiary populations. Among the program's seven criteria of eligibility, three address environmental issues specifically: projects applying for funding cannot promote deforestation or generate any other negative environmental impact; they must include specific environmentally related initiatives and allocate budget for them; and, although only grassroots organizations, such as associations and federations, cooperatives, and unions, may present proposals, environmental NGOs and those supporting indigenous peoples are included as exceptions. The program's requirements for counterpart funds also benefit environmental initiatives.[32] Finally, to avoid dispersion of activities, and to promote the principle of socioeconomic and environmental integration, the program gives priority to funding initiatives by communities located at the margins of major conservation units, Amerindian lands, and rubber tappers' reserves.[33] The underlying principle guiding this rule was to encourage these surrounding communities to develop alternatives to unsustainable practices of exploitation of natural resources in protected areas.

The community initiatives program made sure that Rondonian civil society organizations had direct access to its decision-making processes at several levels. First, by participating in the program's working groups for the analysis and selection of proposals, and monitoring of approved projects. Second, by mandating parity between government and civil society in the program's deliberative council, which decides strategic issues.[34] Third, by establishing a system for transferring funds to program beneficiaries that increased their level of control over these resources.[35]

Despite the detailed environmental provisions of PAIC, some observers argue that it has been merely another vehicle for the state's (and some beneficiary groups') developmental agenda. They even perceive the program's environmental variable as having been, to a certain extent, "forced upon" the program by World Bank officials, who have continued to hope that Planafloro would become an example of a successful "green" project.[36] A different kind of criticism charges the program with diverting attention and resources from civil society away from Planafloro's environmental conditions.[37] While civil society groups focus on the community initiatives program, the state has become free, once again, to default on its commitments to the Planafloro environmental component. These criticisms have been vindicated to some extent, since, in recent years, a conservative administration has reduced the number of forest police and sent proposals to the state legislative to reduce areas of established extractive reserves.[38] It is not clear, at this point, whether or how civil society groups have opposed these measures.

Yet despite criticisms, most representatives of Rondonian civil society organizations perceive the community initiative program as a tool in the process of improving their overall quality of life politically, through the participation opportunities it entailed, and economically, for its direct funding of productive activities. For these individuals, the program's contribution to the environment is an inherent consequence of a larger process of community development.[39]

> [W]e have always pushed for (environmental) sustainable development within the program. Obviously, sustainable development relates to all aspects of a community's life, but it is important to stress that the program should provide incentives to family-based productive activities.... The expectation was that, by obtaining economic sustainability, [Amerindians] would cease to sell wood and to allow mining in their lands.[40]

> What the rubber tappers and the Amerindian communities discuss today is the desire to increase production, but this is within the scope of the program. Yet the communities also want to guarantee the area's environmental preservation. For them it is important to recover areas that have been degraded by logging or colonization.[41]

> I told them that, last year, water had dried out in my plot of land. The only thing that could explain it was that my neighbor and myself had deforested around the head of the stream. Both the forest engineer and the geologist... confirmed that the lack of water was a consequence of us cutting the trees. Put in this way, people easily understood the problem [of the importance of environmental protection for economic productivity].[42]

The processes of formulating and implementing PAIC have contributed to Rondônia's environmentally sustainable development in at least three ways. First, the program has created concrete opportunities—though not always fulfilled—for grassroots communities to formulate their own definition of community development with a sound environmental basis. Second, it has consolidated a process of environmental awareness among populations previously unconcerned with the issue (namely, small farmers). Third, in becoming a symbol of the political strength of Rondonian civil society organizations, the program—and the Planafloro restructuring process as a whole—has created new arenas for cooperation and dialogue between the government and civil society groups on environmental and development policies. The following paragraphs provide details.

The only available evaluation of the community initiatives program to date indicates that its emphasis on "productive" activities has not prevented it from contributing to Rondônia's environment (although to a lesser degree than was originally anticipated). It notes that the apparent dichotomy between environmental and productive activities, which some may see as problematic, may be misleading. For instance, funding for activities such as apiculture and pisciculture may be classified as either "environmental" or "productive." Similarly, activities ordinarily classified as productive (e.g., the recuperation of degraded areas with agroforestry planting) "may deliver a strong long-term benefit for environmental conservation." In conclusion, "many activities and project inputs financed by [PAIC] that are classified as 'productive,' 'infrastructure,' or 'social' are likely to be used by community organizations in ways that promote the environmental conservation goals of Planafloro."[43] In addition, successful projects should stabilize local populations and relieve pressures on conservation areas. It is significant, then, that small farmers—who, in demographic terms, pose the greatest threats to conservation units—were the beneficiaries of 80 percent of projects approved between 1997 and 1998.

The program's contribution to raising the level of environmental awareness among Rondonian grassroots groups, in particular among small farmers, may in the long term, become its major legacy.[44] The role played by key members of the Rondônia network in this process, such as the Rondônia Forum, FETAGRO, and the Rondonian environmental NGO Ecopore, supported by international NGOs with offices in Brazil, such as the World Wildlife Fund for Nature and Oxfam, is undeniable. These organizations have used several strategies to promote the environmental education of Rondonian grassroots groups. To mention just a few, environmental protection became one of the themes in the event *"Grito da Terra,"* ("Cry of the Land"), organized yearly by FETAGRO and MST in Rondônia to discuss issues related to access to land, commercialization of products, and organization of small farmers' cooperatives. The same FETAGRO has established a partnership with the environmental organization Ecopore to initiate an awareness campaign among small farmers on the value of forest reserves. In addition to these specific events, FETAGRO has fostered discussions on environmental issues among its rank and file through the promotion of small meetings in the interior of the state. Anselmo Abreu, president of FETAGRO, however, stresses that there are some differences between environmentalists' and small farmers' perspectives on the environment: "[W]hile environmentalists discuss the issue with an emphasis on the animals, which is correct, we must, in order to convince our bases, address the environmental question from the perspective of the survival of family-based agriculture in Rondônia."[45]

It is interesting to note that FETAGRO has been one of the main forces behind the ongoing efforts of Rondonian civil society to keep the community initiatives program focused on its environmental mission. One of the major threats to the program has been the government's intention to use it to foster the agro-industrialization of the state. Since the end of 1999, Rondonian groups have organized several initiatives to pressure the government to remain accountable to the program's original goals (the state's attempts to undermine the program are discussed in the following section).[46]

The final contribution of PAIC to Rondônia's environmentally sustainable development relates to its having become a symbol of the political strength of Rondônia's civil society. The increased political space that the Rondonian groups obtained during the formulation of the program and Planafloro's restructuring has had concrete effects beyond the scope of these specific processes. For instance, whereas civil society groups did not participate at all in the process of formulating Rondônia's original socioeconomic and environmental zoning plan of the early 1990s, they were a noticeable presence in the zoning plan's second approximation, completed in 2000.[47] The main avenues for civil society participation were the public audiences on the zoning process, promoted by the state in different municipalities, and membership in technical committees in charge of conducting the second approximation. The value of participating in the zoning process was essentially symbolic, since most Rondonian civil society organizations had difficulties coping with its technical aspects.[48] Yet civil society groups occupied a political space that had been traditionally denied to them in the formulation of development and environmental policies for the state.

Opportunities for dialogue and cooperation between selected governmental agencies and civil society groups in Rondônia have also multiplied, as a result of the political empowerment of the latter. For instance, FETAGRO has cooperated with the agency for technical assistance and rural development (EMATER) in agroforestry projects, and CUNPIR has advised state and federal health agencies on Amerindians' priorities. Finally, the cooperation between OSR and the state secretariat for the environment has led to the establishment of infrastructure in the majority of Rondônia's extractive reserves.[49]

This is the good news. The bad news, however, is that the community initiatives program, despite all its promises, to a large extent has fallen pray to Rondônia's traditional politics and to the die-hard habits of political elites of using state's resources for pork barrel schemes. Most important for the argument of this book, is that both the program and Planafloro's environmental component have been compromised by inherent weaknesses of Rondonian civil society groups. These groups have lacked technical, material, and human resources to follow through on their commitments. Ironically, the success of the Rondônia network in establishing significant political space for local civil society participation in the state's development and environmental policy-making also generated burdens that local groups were not able to shoulder. A major question that this book raises, and that will be further discussed in the concluding chapter, is how transnational advocacy networks can best cope with the dilemma that is so well illustrated by the case of the Rondônia network: the success of the network empowers local groups who must then occupy the political space they have conquered. Increased technical and material resources, however, do not necessarily accompany local political empowerment. These deficiencies inevitably reduce the effectiveness of local groups thus threatening their place in the local balance of political forces and, in some cases, even their legitimacy. The following section describes some of the technical and material weaknesses of Rondonian civil society groups, and their consequences both for the implementation of the community initiatives program and for Rondônia's environmentally sustainable development in general.

The Community Initiatives Program:
Political Empowerment versus Technical Challenges

By all accounts, PAIC was an extraordinary achievement. For international observers, the program established a historical precedent.[50] For the representatives of the Rondônia civil society organizations, it was an opportunity for a "more effective participation of civil society in the management of public money."[51] It was also a major step forward in the quest of Rondônia's civil society for political space.[52] Finally, for Rondonian grassroots groups, the program fulfilled many of their expectations;[53] they saw it as one of the best features of the Planafloro project.[54]

It was precisely the high level of demands and expectations that the community initiatives program generated that became one of its curses. The second curse was the intense demand that it created on the technical, financial, and human resources of local civil society groups that participated in it. Finally, the program and the priority placed by local groups on it has alienated the support and interest of national and international members of the Rondônia network. The latter have come to perceive the program as addressing narrow local interests, with little to contribute to larger issues such as environmental and development public policies for Amazonia and tropical forests in general.

Frustrations with the program began at the onset of its implementation. Formal and informal evaluations to date highlight the enormous challenge that the program presented to the limited technical capacity of Rondonian organizations. For instance, the bureaucratic requirements for the approval of funds were difficult to meet. Communities depended upon accredited consultants *("tecnicos")* to draft their proposals according to the program's technical requirements. The processes of selecting and training consultants were cumbersome and often manipulated by the government. Many consultants had low levels of commitment to both the proposals and the communities that they were assisting. As a result, many proposals had identical formats, which were entirely disconnected from the realities of the specific communities they aimed to benefit.

A serious problem that derived from this "industry" of consultants was that some of these individuals encouraged the speedy formation of community associations. Most of these associations formed in response to the opportunities generated by the community initiatives program, rather than to the level of organization and consensus of specific communities. Finally, the program's regional working groups for the selection of proposals (and the representatives of civil society organizations who participated in them) were overburdened by both the number of proposals they had to evaluate and the number of screening processes that had to be conducted. The assessments by Rondonian leaders of the problems with the program (see below) were confirmed by outside observers, who warned about the lack of sustainability of many proposals, an outcome resulting both from the nature of the projects approved, and the level of legitimacy of recipient organizations.[55]

> [The negotiations for the community initiatives program went on until the end of 1996]. After that, the Rondonian civil society had to stretch itself to come up with capable people to draft manuals, define the program's priorities, and its beneficiaries. The effort was to formulate the program's manual in the least complicated format possible. But there must be limits. For instance, the association applying for funds had to be in existence for at least one year. But this condition was waived for Amerindians. On the one hand, such an exception was positive, on the other, it created serious problems. From one minute to the other we went from twelve Amerindian associations in the state of Rondônia in 1994 to thirty-six in 1996–1997. Many of these associations did not even know what their mission was. The FUNAI *tecnico* kept pushing the Indians by saying "you must create the association in order to obtain the [program's] money." This generated an enormous expectation. . . . The tendency [of Amerindian projects] was to prioritize social and economic activities. For instance, fish farming, sustainable agriculture, renovation of schools and health clinics. This was a problem because although these were community needs, the responsibility of providing for them lies with the federal and state governments. But since the governments were absent, Amerindians resorted to the program. What seemed initially to be a lot of money in fact was not, because it was impossible for the associations to replace the government. . . . Also in 1996 CUNPIR assumed responsibility for Planafloro's Amerindian health component. Yet the organization did not have the technical capacity that the project required. For instance, estimates for the payment of personnel were so badly done that in the end CUNPIR could only pay its field staff. There was no money for project management, monitoring, etc. . . . In 1997 CUNPIR drowned in debt and imploded.[56]
>
> [With PAIC] there was a significant progress in defining and conquering an important political space, yet, civil society organization did not have "legs" to conduct the required political monitoring of such a large program. This is a

> problem. Some people back then said that there was cooptation of civil society. I do not think so. Cooptation requires an intention and this was not the case. The government itself was the first to resist the program from the onset. What indeed happened was that many organizations became involved with the execution of the program and in the end, there were not enough human resources to continue to monitor the initiative.... Not only was there a lack of human resources with political training, but also of people with technical knowledge, mostly in the areas of finances and budget.[57]

> Maybe cooptation occurred to a certain extent, but to a large extent, the problem was related to the lack of capacity of these organizations in implementing the community initiatives projects. They were not fully aware of what they were committing to. With all the heavy bureaucratic demands of the program, there was no way they could do anything else. Many organizations were unprepared and had troubles with the accounting part of the projects ...[58]

Local civil society organizations and grassroots groups were clearly challenged by the program's excessive bureaucratic requirements. The application process, for instance, mandated that beneficiaries' organizations present several permits, in itself an expensive process. By the time working groups evaluated the proposals the original permits were frequently outdated. The proposal was then rejected and the communities had to reapply for the permits, pay fees again, and hope for a timely reevaluation of their application. After the approval of a proposal, the government systematically delayed transferring of funds. This often created insurmountable operational problems for the beneficiaries, who were already short on cash. To illustrate, there was a complete disregard from the government/Planafloro bureaucracy for the agricultural calendar (planting season, rainy season, harvest). By the time funds were available, organizations had to struggle to spend the money in activities other than those they had originally planned for.

Besides all the ongoing technical challenges that the community initiatives program imposed on the Rondonian groups, they also had to deal with illegal behavior on the part of the state in managing the program's accounts. 1998 was an electoral year and the outgoing Raupp administration campaigned for reelection. In a desperate move to gain the support of civil servants, whose salaries had been withheld due to lack of public funds, Governor Raupp arbitrarily (and illegally) diverted the money from the community initiatives program to the state's payroll.[59] The consequences were dire for community associations and NGOs that had already committed those resources. Projects were paralyzed, crops were lost, and frustrated suppliers initiated lawsuits against associations that failed to pay for preordered products. Rondonian civil society representatives cited examples of some of the consequences of the government's "betrayal":

[P]eople were in real trouble, a leader of one association committed suicide, other groups had to organize raffle schemes to cover their expenses . . .[60]

The small farmers, who had ordered seeds and plants [were very affected]. Several associations of small farmers are being sued by suppliers who have not been paid. . . . Delay in transferring funds is harming one community initiative program proposed by the rubber tappers that aimed at building infrastructure for eco-tourism in the entire state. We [the rubber tappers] expected to have it in place a year ago. In June [2000] we will host groups of international tourists and our infrastructure is limited to a small hut, built by the community itself.[61]

While the crisis generated by Raupp's maneuver failed to reelect him, it succeeded in paralyzing the community initiatives program for the entire year of 1999. The conservative administration that replaced Raupp's in 1999 showed no interest in resolving the crisis of the community initiatives program. It was only through the pressures of civil society groups and of the World Bank Planafloro team that a deal was eventually negotiated between the bank, the Brazilian federal government, and the state of Rondônia. Rondônia borrowed money from the federal government and the community initiatives program was reinitiated in 2000. The resolution of financial issues was not the only determinant of the program's reinstatement. The Rondonian government has become increasingly aware of the political opportunities that such a program entails and has attempted to increase its control over processes of resources allocation. This presents two dangers: first, the program may become another instrument for pork barrel politics; second, it may lose its environmental focus and revert to an emphasis on productive activities for short-term results. Although Rondonian civil society groups are committed to preventing the government's "takeover" of the community initiatives program, it is not clear whether they will succeed.

Taking Stock of the Rondônia Network at the Turn of the Millennium

The overcoming of the 1994 legitimacy crisis marked the political maturity of the Rondônia network's local membership base. In fact, it indicated the completion of a process of "localizing" the network's activism. In reasserting their commitment to the interests of their local constituencies, Rondonian civil society groups demonstrated their increased autonomy vis-à-vis their international partners. This new phase, characterized by a higher level of parity among network members, created fertile conditions for the strategy of demanding the investigation of Planafloro by the inspection panel. The Planafloro inspection panel claim at once generated and made evident an unprecedented level of cohesion among network members. The political struggles that it triggered, both in international arenas (the World Bank board of directors) and in Rondônia, to a certain extent responded to the concurrent goals of network members of forcing changes in Planafloro's implementation and testing the effectiveness of the panel as an instrument of World Bank accountability. In sum, the claim was successful on three counts: first, it became a catalyst for the harmonization of the interests of the international and domestic members of the network vis-à-vis both Planafloro and the larger environmental policies of multilateral development banks. Second, it was an effective tool in promoting the cohesion of network members. Third, it was a powerful strategy, whose visibility and novelty lent to local groups an unprecedented level of political leverage within Rondonian politics.

Examples of the increased political leverage of Rondonian civil society organizations proliferated, particularly immediately following the filing of the Planafloro inspection request. The most significant of such examples was, without dispute, the restructuring of the Planafloro project and the formulation of the community initiatives program. Members of the Rondônia network themselves best describe the actual and symbolic significance of the program:

> In my evaluation, the community initiatives program was a complete and very positive experience, despite its flaws. Sure, there were many badly implemented projects, projects that lacked a participatory approach, there was a lot of internal confusion within the communities, but there are also many projects in the field that would not have been there were it not for the program. . . . Unfortunately, in the long term, I think that the continuity of the program is a remote possibility given the state's political and economic situation . . .[62]

> The program is thus a way of transferring resources from the state to civil society organizations that are most permeable to environmental issues. The project is important because it promotes a more decentralized style of management and a more sustainable agriculture. In the case of the rubber tappers, the program even had a more direct impact in the extractive reserves, and in the case of Amerindians, it has encouraged alternative activities to logging and mining.[63]

Despite its merits, the strategy of filing the Planafloro inspection panel claim also revealed, in the long term, at least three remaining weaknesses of the Rondônia network. First, the political empowerment experienced by local groups as a result of the claim was not accompanied by an equivalent increase in their technical capacities. As a result, local groups stretched themselves too thin, trying to fill the political space they had carved. Despite some significant successes, local civil society groups have been unable to effectively confront the backlash that the Rondonian conservative forces have unleashed since 1998 against many of the gains of Planafloro.

The second weakness of the Rondônia network is the continuing presence of divisions among local groups. The tensions deriving from their limited technical capacity and from the backlash of the Rondonian elites against Planafloro, have reopened old disagreements about the role of local groups vis-à-vis the project. For instance, since former governor Raupp diverted funds from the community initiatives program, and the incumbent administration attempted to redirect the program toward unsustainable productive activities, the landless movement and the federation of agricultural workers of Rondônia have withdrawn from the program. This means that while agricultural cooperatives and rural associations may continue to apply for community initiatives funding, their leading organizations no longer participate in the program's joint implementation and decision-making mechanisms. Rubber tappers and Amerindian groups, however, have continued to participate in the program, thus legitimating it despite its problems.

Another interesting aspect of the divisions among Rondonian groups is the relative weakening of the Rondônia Forum's role as a catalyst of civil society's environmental activism in Rondônia. Although this is a recent phenomenon and merits further analysis, the decline of the forum's role in Rondônia politics may be explained by several factors, not all of them detrimental to the political future of Rondonian groups. It is evident that divisions among its support and member organizations contributed to diminish the forum's effectiveness. Yet the most serious factor contributing to the forum's weakness was its lack of capacity. The forum was hurt by the burnout of many of its leaders during the intense processes of filing the inspection panel claim and negotiating the community initiatives program. The very success of these strategies opened the way for these individuals to advance their careers elsewhere, leaving a vacuum of technical expertise in the forum. This happened at the same moment that the organization's commitments of co-implementing and monitoring both Planafloro and the community initiatives program increased. On a brighter note, however, part of the explanation for the forum's diminished political role in Rondônia relates to the growing role, capacity, and autonomy that other civil society groups have achieved in the last ten years. Whereas in 1991, when the forum was formed, many grassroots groups such as the Organization of Rondonian Rubber Tappers were in their early infancy, and others such as CUNPIR did not even exist, in the twenty-first century these groups and others such as FETAGRO are fully established in Rondônia. Increasingly, they have found their own voices and made them heard in Rondonian politics, independently of a formal mechanism originally created to amplify their collective voices. In transitioning from the forefront to the background of local socioenvironmental activism, the Rondônia Forum may encounter a new vocation, one of providing support to the initiatives of civil society groups rather than initiating or being the catalyst of such activism.

Finally, the third weakness of the Rondônia network reflects a recurrent problem with transnational advocacy networks. It refers to the fragility of Northern-Southern coalitions.[64] In the case of the Rondônia network, the inspection panel claim in particular revealed the difficulty that transnational networks face in sustaining mobilization and high levels of cooperation and partnership among their members in the long term. For international NGOs, the rejection of the Planafloro claim represented the end of a process. The claim tested the panel and triggered efforts for its reevaluation and restructuring as an international mechanism for development agencies' accountability. These processes then became the new focus of interest for many international groups who were once involved in the Rondônia network. For the Rondonian groups, however, the claim was the start of a new phase in their ongoing struggle for participation in the state's environmental and development decision-making processes. This is a struggle that the Rondônia groups have fought alone, particularly since the formulation of the community initiatives program in 1996.

Quite possibly, the key element in processes of institutionalizing Northern-Southern advocacy coalitions is the mediation of national organizations or networks of organizations. This link has been chronically missing in the case of the Rondônia network. Historically, Rondonian groups have had direct access to international NGOs (and vice versa) due in large part to the personal connections established among leaders in Rondônia and abroad dating from the time of the Polonoroeste mobilization. Since then, however, Brazil has democratized and its civil society has been actively devising mechanisms to increase popular participation in policymaking. Successful examples of this effort include the creation, in 1995, of a national coordinating body of organizations concerned with the policies of multilateral development banks in Brazil (the Brazil Network on Multilateral Development Banks, or Rede Brasil), and collective bodies such as the Brazilian National Forum of NGOs and Social Movements on the Environment and Development. Rondonian organizations, however, have been noticeably absent from these mechanisms. The challenges of Rondonian politics and the overstretching of local organizations due to their commitments within Planafloro may explain their incapacity to establish stronger links with national advocacy entities. Yet this very incapacity has further compromised their chances of advancing technically and politically.

International NGOs, eager to assert their transnational legitimacy on the basis of their alliances with local groups, have bypassed, often inadvertently, national clearinghouses and collective mechanisms. As a consequence, they have missed opportunities to enlarge the scope of their international activism in terms of the long-term goals of national civil societies. Unless international groups, often the most resourceful elements of transnational advocacy networks, commit such resources to long-term, incremental struggles, transnational advocacy efforts will remain limited to the sporadic successes of specific strategies and campaigns.

In light of the above discussion on the issues that hinder the effectiveness of transnational environmental advocacy networks, I return to the question raised at the onset of this chapter: what is the relation between transnational environmental activism and local political struggles for citizenship rights? It is my understanding that the answer to this question depends on how activists approach the concept of environmentally sustainable development. If the approach is a "mainstream" or "conservationist" one, that is, environmental protection is defined strictly in terms of the protection and conservation of natural resources, the relationship between transnational environmental activism and local struggles for citizenship is a limited one. At best, transnational environmental activism will contribute to promote limited citizenship rights directly related to environmental protection. It should be noted that in the developing world, immediate survival often overshadows goals of environmental preservation in a strict sense.

If the members of transnational environmental advocacy networks, however, approach environmentally sustainable development as an integral process of improving the quality of life of local people, environmentally, politically, and socioeconomically, then the relation between transnational environmental activism and local struggles for citizenship becomes a strong one. As illustrated by the process of "localizing" the Rondônia network, the pursuit of environmentally sustainable development in the state quickly became an all-encompassing effort toward the improvement of the quality of life of sectors of the local population. That meant not only creating conservation units and Amerindian reserves and maintaining their integrity, but also affecting legislation and policies detrimental to natural resources conservation. In addition, it meant striving for direct participation of civil society organizations in arenas that were once under complete control of the state and local elites (such as the formulation and implementation of environmental and development public policies). Finally, it meant devising environmentally sound alternatives for the improvement of social and economic conditions of local populations as counterweights to unsustainable practices.

Here the three-dimensional nature of transnational environmental networks becomes a complicating factor. To what extent are international members of the network capable or willing to participate, even if indirectly, in local struggles for citizenship rights? To what extent can local citizenship struggles be fully subsumed into national struggles? How much re-framing of local demands may be necessary in order to include them in the agendas of national organizations, such as political parties and umbrella civil society organizations or clearinghouses? And, last but not least, to what extent can national struggles for citizenship rights contribute to the immediate and specific needs of local groups? If transnational environmental advocacy networks are to have long-term impacts on local human and natural environments their members must address these questions. The investigation of transnational environmental activism presented in the following chapters, against oil exploitation in Ecuadorian Amazon and dam construction in India, gives further strength to this proposition.

6

Environmental Activism beyond Brazil I— The Struggle against Oil Exploitation in Ecuador

The analysis of the Rondônia network raised important issues and illuminated significant dynamics of the internal politics of transnational advocacy environmental networks. The present chapter and the one that follows provide an opportunity to contrast these issues and dynamics with those of other transnational advocacy networks, namely, the one fighting oil exploitation in Ecuador's Amazon region (known as the *Oriente*) and the one resisting the construction of the Sardar Sarovar dam on India's Narmada river (see chapter 7).

The selection of these two cases followed the same theoretical rationale that elected the Rondônia network as the main case study for this book: the actions of both the anti-oil network and the Narmada network obtained a high degree of visibility in the late 1980s and early 1990s, and became landmark cases for environmental and human rights advocacy throughout the world. In addition, they both evolved during a period of more than fifteen years (from the mid-1980s to date), thus allowing for the identification and analysis of their structural characteristics. The fact that they coincide in time with the Rondônia network also facilitates comparison by allowing a certain degree of control of historical variances.

The study of the anti-oil network expands the scope of the analysis of transnational advocacy networks in a new direction. It brings the private sector (i.e., multinational oil corporations) into the discussion of transnational environmental activism. It allows for a comparison between activism that confronts state institutions at several levels and activism directed, primarily but not exclusively, at specific companies and/or a specific industry.

Fighting Oil Exploitation in Ecuador's *Oriente:* Background and Network Origins

Since the early 1960s, the upper Ecuadorian Amazon region known as *Oriente* has been opened to oil exploitation. Around ten to twelve multinational oil giants have operations in the *Oriente,* as well as Ecuador's own Petroecuador. Oil prospecting and exploitation require procedures, such as dynamite use, the drilling of wells, and construction of heliports and pipelines in the jungle, that are highly detrimental to the stability of ecosystems. The pipelines in particular are susceptible to all sorts of accidents, and given the remoteness of the jungle and the absence of communication and operational infrastructure that could facilitate emergency action, such accidents often result in large oil spills. Yet, while oil operations per se have significant impacts on the stability of ecosystems, processes of development and colonization of the jungle that follow in their wake have even worse effects. Oil operations require roads for the transportation of equipment and personnel. Once roads cut through the forest, they open the way for masses of landless migrants who, in an impoverished country such as Ecuador, see unclaimed land in the jungle as their last hope to make a living. Slash and burn practices prevail and soon the forest disappears. Both settlers and oil companies encroach on Amerindian territories, disrupting their inhabitants' way of life and endangering their physical integrity.[1]

Historically, however, neither environmental nor indigenous rights have been objects of serious consideration by the main actors shaping Ecuadorian economic policies. Modern policymaking in the country is constrained by two legacies. On the one hand, there is an authoritarian tradition that has permeated both civilian and military governments, and has contributed to the perpetuation of the executive branch of government as the dominant power. On the other hand, there is the size and the power of the state bureaucracy. In the case of economic development and energy policies, for instance, the role of Petroecuador, the state-owned oil company, is determinant. In the 1970s, high oil prices allowed the government to increase spending and the external debt. In the 1980s and 1990s, however, low oil prices led the country to an economic crisis from which it has yet to recover.

As it happened in most Latin American economies during that period, Ecuador responded to its economic difficulties by implementing structural adjustment policies. These policies received the technical and financial support of both the World Bank and the International Monetary Fund. Structural adjustment in Ecuador meant, among other things, encouraging exports in the oil and agricultural sectors, and initiating steps toward the privatization of key state-owned companies, chief among them Petroecuador. The last thing that the Ecuadorian government wanted in the midst of these processes was an international campaign that highlighted the environmental and human risks of oil exploration in the country. This was exactly what it confronted in the early 1990s.

The book *Amazon Crude,* by the scientist and activist Judith Kimerling, was among the first to document the impact of oil exploitation on the human and natural environments in the *Oriente.* The book focused on the operations of Texaco, and revealed that by the (conservative) estimates of the Ecuadorian government, the company had spilled 16.8 million gallons of crude into the region's rivers—one and one-half times the amount released by the *Exxon Valdez*—and dumped 19 billion gallons of toxic waste waters into the environment.[2] Since the publication of the book, the environmental and social consequences of oil exploitation in Ecuador have been further investigated, and resisted, by a coalition of international environmental and human rights organizations, environmental and human rights groups in Ecuador, Ecuadorian indigenous peoples' organizations, federations, and confederations, independent activists, sectors of the national and international specialized media, and most recently, select members of Ecuador's political elite. This coalition operates at several levels: it has targeted World Bank loans for Ecuador's oil sector,[3] it has been active in influencing the country's indigenous rights and environmental legislation,[4] and it has mounted campaigns against the environmentally unsound practices of Petroecuador, as well as specific multinational corporations.

To pursue the comparative objectives of this chapter, I chose to discuss two specific instances of activism of Ecuador's anti-oil transnational environmental advocacy network: the campaigns against the oil giants Texaco and ARCO Oriente (a subsidiary of the U.S.-based Atlantic Richfield). The main reason for thus narrowing the analysis is to be able to identify specific local groups most engaged in these campaigns, and assess the consequences of such an engagement for their political and technical empowerment. In addition, the focus on specific campaigns should facilitate the assessment of their consequences for the protection of the local environment.

Any study of transnational environmental activism in Ecuador must be mindful of the atmosphere of distrust that constrained original attempts to establish local, national, and international cooperation on environmental and indigenous issues. Between 1990 and 1991, a controversy emerged in U.S. newspapers and magazines about the role of international environmental NGOs, particularly the Natural Resources Defense Council (NRDC), in mediating negotiations between Ecuadorian indigenous groups and oil companies. There were charges that the NRDC had attempted to broker a deal between the U.S. oil company Conoco and the Huaorani indigenous group.[5] The deal would guarantee that a portion of profits would fund independent monitoring of Conoco's oil operations and provide resources for indigenous peoples' projects. Environmental groups in Ecuador as well as some U.S. NGOs vehemently condemned NRDC for considering Conoco's proposal in the absence of a formal mandate from Ecuadorian indigenous groups. In its defense, NRDC maintained that it had received a request from CONFENIAE (Confederation of Indian Nations of Ecuadorian Amazon) for support in negotiations with oil companies. Shortly after the controversy became public, NRDC interrupted its work in Ecuador.

The Conoco/NRDC controversy highlighted potential divisions not only between international and national/local groups, but also between Ecuador's young environmental movement and indigenous organizations.[6] While environmental groups opposed oil exploitation on principle, indigenous peoples were divided on the matter, and some groups welcomed the opportunity of obtaining direct material gains from the oil companies. The Conoco/NRDC controversy and what it revealed about the complexity of relations between environmental and indigenous groups in Ecuador, as well as within different indigenous organizations, strongly influenced the type of network that was later established between international and national/local groups on oil issues in the *Oriente*. They highlighted the importance of trust among network members, and of asserting the legitimacy and accountability of one's organization vis-à-vis both its constituency and other network members.

Thus, as international environmental groups became increasingly interested in the environmental impacts of oil operations in Ecuador and initiated international awareness campaigns on the issue, one paramount concern was that the legitimacy of their initiatives was not undermined by lack of consultation with local groups. In the early 1990s popular mobilization was not a new phenomenon in the *Oriente*. Since the 1970s both Amerindian groups and settlers had started to organize around land issues. For Amerindian groups in particular, asserting their territorial rights was a matter of survival since both oil companies and settlers increasingly encroached on their lands. To best fight for official recognition of their territorial rights, Amerindian groups formed organizations that were usually based on specific ethnic identities. Texaco operations and the development processes that came in their wake directly affected the Cofan, Secoya, Siona, and Quichua Amerindian groups, who formed organizations identified by their initials, respectively OINCE, OISE, ONISE, and FCUNAE. In the area in which ARCO originally operated (known as Block 10), the affected indigenous groups had come together under the Organization of Indigenous Peoples of Pastaza (OPIP). During the 1980s, the small ethnic-based organizations quickly moved to form regional, and later national, organizations. The *Oriente* Amerindian organizations joined together in CONFENIAE and, in 1986, CONFENIAE and the highlands Amerindian groups formed the National Confederation of Indigenous Nationalities of Ecuador (CONAIE). The

Catholic Church and local human rights groups were key supporters of Amerindian groups in their organizational effort.[7]

While the environmental and health impacts of oil operations, and those of Texaco in particular, did not originally motivate Amerindian and settlers' mobilization (access to land was the primary issue), they quickly became objects of concern, thanks to the awareness efforts of both church-based groups and individual activists such as Kimerling. Since 1987, studies by the Ecuadorian government and by independent sources had disclosed information on Texaco's environmental "legacy." Besides the oil spills and the toxic production waters that were discharged directly into waterways and groundwater sources, the company had unlined and uncovered waste pits throughout the *Oriente,* and was directly or indirectly responsible for the deforestation of more than two million acres of forests.[8] In 1990, Ecuadorian environmental groups led by *Acción Ecologica* (Ecological Action) held a meeting with Amerindian and settlers' organizations in the *Oriente* and launched the "National Campaign Amazon for Life" *("Amazonia por la Vida")*. The goal of the campaign was to protect the *Oriente's* environment and its population, particularly Amerindians, from the effects of oil exploitation. The campaign's main focus quickly became Texaco's irresponsible environmental practices in Ecuador. Activists demanded that the company conduct clean-up operations and be accountable to those populations affected by environmental contamination. Key international supporters of the campaign were the U.S.-based NGOs Oxfam America, Rainforest Action Network (RAN), and the Center for Economic and Social Rights (CESR). In time, international membership in the network was greatly expanded through the work of the coordinating office of the Amazon Coalition (which I discuss in the following section).

The history of the anti-oil network coincides, in part, with the evolution and strengthening of the Ecuadorian indigenous movement. This history is extraordinary for the movement's mobilization capacity, diversity of strategies, and effective conquests. In the last twenty years, Ecuadorian indigenous peoples have succeeded in securing a voice in the political system through intense and well-coordinated activism.[9] While their actions are not analyzed in detail in this book, it is important to keep in mind that they affected the political context within which the anti-oil transnational advocacy network, and its campaigns against Texaco and ARCO, unfolded.

Indigenous peoples in Ecuador were, until the mid-1990s, among the most excluded sectors of the national society. Economically, they are heavily dependent on subsistence agriculture and have been negatively affected by structural adjustment reforms that aimed at modernizing the agricultural sector and geared it to export markets. Socially and politically, indigenous peoples' participation has been hindered by a Hispanic political culture that rejects traditions and practices considered incompatible to the modern nation-state.[10]

Since the 1970s, regional and national indigenous peoples' organizations have voiced two main demands: access to land and to bilingual education (the latter is considered a key instrument in the process of consolidating and valuing an indigenous identity). In 1989, CONAIE and the minister of Education and Culture finally signed an agreement for the establishment of an agency to oversee bilingual schools throughout the country. The agreement was a result of more than twenty years of campaigns for bilingual literacy programs and of the intensification of the dialogue between the government of president Rodrigo Borja (1988–1992) and indigenous organizations.[11] Besides its obvious importance as a resource for bilingual schools, the agency also had a symbolic importance. It indicated the government's recognition of an indigenous identity that is specific to certain sectors of the Ecuadorian population.

Symbolic advances, however, did not address the demands by indigenous peoples for land regularization. In July 1990, CONAIE organized the first national indigenous uprising, which aimed at pressuring the government to address land conflicts in the country. Such conflicts, many of which had remained unsolved despite agrarian reform laws issued in the 1960s and 1970s, intensified in the 1980s. The uprising was also a reaction against the government's complete neglect of indigenous peoples' interests in the structural adjustment initiatives for the agricultural sector.[12]

The main consequence of the 1990 uprising was the strengthening of indigenous peoples' voices in Ecuador's political system. From the perspective of land regularization, however, success was much more limited. During the early 1990s, the Ecuadorian government intensified negotiation with multilateral finance institutions to obtain loans for the modernization of the agricultural sector. Negotiations led to the 1994 Agricultural Development Law, which aimed at the liberalization of the land market. The law undermined at its core the indigenous land management system, based on the collective ownership of the land. After several unsuccessful attempts to influence the drafting of the Agricultural Development Law, CONAIE called for a second uprising to demand that the law be repealed. The series of mobilization initiatives that culminated with hundreds of thousands of indigenous peoples blocking the Panamerican Highway, and thus the transportation of agricultural products to urban centers, was labeled "Mobilization for Life."[13] The label seems to indicate a clear connection between the mobilization against the reform of the land sector and the Amazon for Life campaign, which among other things, opposed international loans to increase oil exploitation in Ecuador (the Amazon for Life campaign is discussed in further detail later in this chapter).

The 1994 uprising contributed to further increase the power of indigenous people in Ecuador's politics. It also had concrete impacts on both the agricultural law and on the policies of the Interamerican Development Bank (IDB) for the country.[14] After several episodes of repression and intimidation of indigenous organizers and their allies, the government of President Durán Ballen agreed to negotiate. Meetings between government officials, representatives of indigenous peoples, and members of other civil society organizations led to revisions in the agricultural law. CONAIE was also able to negotiate measures to mitigate the impact of the IDB loan to the agricultural sector.

The strength of the indigenous movement in Ecuador reached unprecedented levels by the late 1990s. Sectors of the indigenous movement have formed the Pachakutik political movement, which provides resources for campaigns of indigenous candidates to representative positions. Several indigenous representatives were elected to participate in the 1998 National Assembly. They became instrumental in reforming the constitution, which now defines Ecuador as a plurinational state. Most recently, in January 2000, street demonstrations led mainly by indigenous peoples forced President Jamil Mahuad out of office. While the military eventually assumed the upper hand in processes that followed in the wake of Mahuad's departure the Ecuadorian indigenous movement has entered the new millennium aware of its political strength and with an unparalleled "sense of possibility."[15]

Weaknesses and Strengths of the Anti-Oil Network

The previous section listed the main actors involved in the anti-oil transnational advocacy network in Ecuador and traced their evolution toward the launching of the campaign against Texaco. Before I proceed with the analysis of their strategies and how the network and the campaign affected local groups in the *Oriente*, I shall discuss in further detail two specific aspects of the anti-oil network that had a direct effect on the level of empowerment of its local members. First, it is important to highlight the fragile nature of the alliance between actors whose agendas often contradicted each other, as in the case of leading Ecuadorian environmental NGOs, Amerindian organizations, and settlers' groups. The second distinguishing aspect of the anti-oil network refers to the role played by the Amazon Coalition.[16] The nature of the coalition and the way it operated became an unprecedented resource for the network.

Network Cleavages. Analysts testify to the uneasy relationship between environmental and Amerindian groups in Latin America and in Ecuador in particular.[17] The different priorities in these groups' agendas and their different modes of operations are responsible for the main divisions. For indigenous groups, land rights and autonomy/sovereignty are the primary concerns. Environmental groups tend to focus on conservation efforts and on the establishment of protected areas or ecosystems. While this discrepancy exists even among activist organizations in North America, they are particularly striking in Latin America.[18] They definitely affected the anti-oil network in Ecuador. For instance, one of the difficulties in reconciling the interests of indigenous peoples and environmentalists (in Ecuador and abroad) relates to the need to develop proposals that would allow, simultaneously, the sustainable management of natural resources and the satisfaction of indigenous populations' needs for subsistence and market goods.[19]

Another difficulty of Amerindian participation in an "environmental" network is the issue of autonomy and representation. The strength of the indigenous movement in Ecuador has been based, among other things, on its capacity to create large and representative umbrella organizations. Despite the inevitable disagreements that occur among nationalities and individuals within indigenous confederations such as CONFENIAE and CONAIE, overall indigenous unity has prevailed. Ecuadorian indigenous peoples are proud of their organizations and resist having their interests "represented" or voiced by any other group or coalition. As Jezic explains when discussing some of the strategies of the anti-oil network against Texaco, "Only indigenous organizations can speak for indigenous peoples. . . ."[20] The same idea is corroborated by Selverston-Scher when she explains that, in part, the Amazon Coalition was an effort by indigenous peoples to control outside (i.e., Northern) activism that bypassed them.[21]

Finally, if an alliance between environmentalist groups inside and outside Ecuador, and Amerindian groups was complicated by the reasons discussed above, the inclusion of settlers' organization was an even greater challenge. Despite their goals of material improvement, sometimes through the exploitation of natural resources within their lands, indigenous peoples perceive themselves as living in harmony with nature.[22] The threat to their sustainable way of life comes from colonists, as well as from large businesses.[23] As discussed in previous chapters, similar divisions opposed environmentalist and indigenous peoples' organizations, and landless and small farmers' associations in Rondônia at the onset of the network's activism.

In light of these cleavages, how was the anti-oil network successful in bringing these groups together in the same mobilization effort? Two key factors—and the individual efforts of specific activists—account for the answer. The first basis for common ground among environmentalists, indigenous peoples, and Amazon settlers is that oil exploitation has affected all of them, albeit at different levels. Environmentalists are concerned with the destruction of the forest and the loss of ecosystems and biodiversity. Indigenous peoples have been affected by the presence of oil operations (drilling wells, seismic activities) and oil companies' employees in their lands (often without previous warning or authorization by indigenous groups) Operations have affected hunting grounds and burial sites, and the proximity of non-Indians has brought diseases and disrupted cultural traditions. Last but not least, oil spills and poisonous discharges have directly affected the health of indigenous peoples and settlers alike.

The second factor concerns resources. In the late 1980s it became evident that heightened environmental concerns throughout the world offered an opportunity for local groups to tap into the resources of international environmental NGOs. In the case of Ecuador, national environmental NGOs such as *Acción Ecologica* and CORDAVI *(Corporacion de la Defensa de la Vida)* had close contacts with international groups and thus tended to be the primary mediators or coordinators of actions between activist groups at the local level and their potential international supporters. Environmental organizations both in Ecuador and abroad also played a significant role in producing information about oil operations and disseminating it among affected groups. This contributed to raising awareness about the common threat affecting populations in the *Oriente* and the need for coordinating their resistance. Later in this chapter I discuss how specific strategies contributed to—and sometimes hindered—the strength of the alliance between network members in their struggle against Texaco (and later, against ARCO).

Network Resources. It is important to highlight the role of the Amazon Coalition as a particularly interesting player in the anti-oil network. In the early 1990s the coalition became a major source of political and technical resources to national and local network members. In fact, its nature and organizational structure contributed to set apart the anti-oil network from the other networks discussed in this book. The Amazon Coalition made possible an unprecedented level of trust among national/local and international network members. It is important to remember the context that preceded the creation of the Amazon Coalition in 1990. In Ecuador in particular, national and local environmental and Amerindian groups had become wary of international NGOs' activism following the Conoco/NRDC controversy. Throughout Latin America one could sense a similar atmosphere. Selverston-Scher thus describes the mood among the members of the Coordinating Body of Indigenous Organizations of the Amazon Basin (COICA), an organization that attempts to coordinate the interests of a large number of indigenous groups in the countries of the Amazon Basin:

> [T]hey felt that US environmental groups were tromping around their rainforest homeland without taking indigenous inhabitants into account. The extreme problems were setting up protected areas and kicking the Indians out of them. The more frequent problems were a general lack of consultation or respect, particularly of Indigenous organizations.[24]

In order to confront these problems, in 1990 COICA called for a summit between international and Latin American environmentalist and indigenous groups in Iquitos, Peru. The creation of the Amazon Coalition was one of the results of this meeting. Without going into the specific details of the coalition's organizational and decision-making structures, it is important to highlight three of its features that greatly facilitated the actions of the anti-oil network. First, the Amazon Coalition was designed to be very flexible in its formal membership requirements and organizational structure. Second, it aimed at being as broad and inclusive as possible, embracing indigenous organizations of different sizes and political stature, such as NGOs, "support organizations" (research institutes and advocacy groups), environmental organizations, and the organizations of "other" forest peoples, such as maroons and rubber tappers. Third, the coalition's main role was that of a mediator and facilitator of interorganizational cooperation. In that, it strived to guarantee that all coalition members benefited from its actions and pursuits, even if not in similar ways (in a given action, for instance, some members would benefit from the political visibility they obtained, or from asserting their legitimacy in a context or struggle, others would benefit from an increased technical capacity, such as training, knowledge, and access to information, and finally, some members would benefit from direct access to financial/material resources channeled by the coalition).[25]

As I discuss the specific strategies of the anti-oil network against Texaco and ARCO, the contribution of the Amazon Coalition to the network's unity and international visibility will become evident.

THE ANTI-OIL NETWORK AND THE CAMPAIGNS AGAINST TEXACO AND ARCO

The Campaign against Texaco

Within the context of the broader mobilization against oil operations in the *Oriente* (the Amazon for Life campaign), the starting point of the campaign against Texaco was the occupation of the company's office in Ecuador's capital, Quito, on June 28, 1991. The move was conceived and organized by the environmental NGO *Acción Ecologica*. Domestic pressures focused on the Ecuadorian government, demanding that it evaluate the environmental impact of Texaco's operations and mandate cleanup initiatives. These pressures were coordinated with international calls for a letter-writing campaign and a Texaco boycott.[26] Between 1991 and 1993 the campaign against Texaco divided its resources among the pursuit of several strategies, such as demonstrations, marches, occupation of Texaco's and the government's offices, media campaigns, organization of fact-finding missions by Ecuador's congressional representatives to the *Oriente,* and an inordinate number of meetings with government and industry officials.[27] Except for raising awareness within and outside Ecuador about the environmental and social problems caused by oil activities in the Ecuadorian Amazon, these actions accomplished little.[28]

The domestic and international activism against Texaco captured the attention of a lawyer, Cristobal Bonifaz, who, with the support of the New York-based Center for Economic and Social Rights, filed a lawsuit against Texaco in a United States District Court.[29] The lawsuit represented a groundbreaking strategy for the anti-oil network for several reasons. First, it was instrumental in strengthening the fragile alliance between settlers' and indigenous organizations representing the populations directly affected by Texaco operations in the *Oriente* (the lawsuit was conceived as a class-action suit on behalf of all affected residents, around thirty thousand individuals, without discriminating among them). Second, it created a situation in which the government of Ecuador was forced to adopt a formal position vis-à-vis the charges against Texaco. This occurred in light of successive requests by the U.S. courts for evidence that the Ecuadorian government supported the plaintiffs' claims and did not oppose U.S. jurisdiction in the matter. Third, it raised the possibility of "changing the course of legal history,"[30] if it were to set a legal precedent by which multinational corporations might be held accountable for the consequences of their operations anywhere in the world. The argument of the plaintiffs' lawyers is that the decisions that harmed the *Oriente* and its populations were made at Texaco's U.S. headquarters. Thus, documentation about such decisions would only be made available to a U.S. court. Not surprisingly, Texaco has both denied the charges and insisted that the case be tried in Ecuador.

There is extensive information available about the Texaco lawsuit, the legal debates it has triggered, and its political implications for Ecuador's sovereignty and the future of the oil industry in the country.[31] In accordance with the stated objective of this chapter, however, I will limit my discussion to the impact of the lawsuit and other related strategies on the local organizations participating in the anti-oil campaign and on their natural environment.

One major challenge that members of the anti-oil network have had to face when confronting specific companies is the attempts made by the latter to disrupt network consensus and unity. As discussed above, different priorities have characterized the relations among network members from the onset of mobilizations. As the campaigns against oil companies have unfolded, the network has continued to face risks of internal division that originate both from within and outside the network. In the case of the Texaco campaign, there were differences among network members about the legitimacy of pursuing a lawsuit in foreign courts. For some it clearly undermined the country's national sovereignty and the credibility of its legal system. Most damaging, however, were Texaco's attempt to break activists' unity by offering compensatory deals to specific indigenous groups and/or organizations. To a significant degree, the mediation of the Amazon Coalition at specific junctures contributed to ease some of the effects that these threats posed to the network's national-international connection.

For instance, at the request of Ecuadorian groups, the Amazon Coalition played a crucial role in coordinating international initiatives to alter the position of the incoming administration of President Jamil Mahuad vis-à-vis the lawsuit against Texaco.[32] While a previous administration had supported the plaintiffs, Mahuad's indicated to the New York courts, in 1998, that it no longer considered Texaco responsible for environmental damages in the *Oriente*. The Amazon Coalition coordinated the efforts of the international members of the anti-oil network in support of domestic activism to alter the position of the Mahuad administration.[33] The Amazon Coalition counted on two major assets upon which to base its participation in this process. First, different from other initiatives of transnational advocacy networks unfolding in international arenas, those implemented by the coalition could not be perceived, either in Ecuador or elsewhere, as driven by "foreign organizations." The coalition's diverse membership, in geographic and institutional terms, asserted its identity as an organization of the "international civil society."[34] Second, its lobbying efforts—a letter-writing campaign and media mobilization—were supported by the argument that international groups, like Ecuadorian organizations, had a legitimate stake in the continuation of the Texaco lawsuit in a foreign court. According to the coalition, the demand is a "precedent-setting case with ramifications for countries and communities throughout the Amazon and the rest of the world."[35]

Besides helping to coordinate the efforts to sustain the Texaco lawsuit in the U.S. courts, the Amazon Coalition also played a role in strengthening the alliance among local groups in Ecuador. Starting in 1995, it co-sponsored several workshops where participants were encouraged to devise regional strategies to defend the Amazon against oil development. The workshops were open to representatives of NGOs, indigenous federations, and communities affected by oil development. They were organized around training sessions that focused on communities' legal rights in face of oil activities, and on opportunities for exchange of information and experiences that could lead to further regional campaigns of resistance.[36]

At the local level, there were two important consequences of the Texaco campaign and of the awareness efforts of network members: the creation of the *Frente de Defensa de la Amazonia* (Amazon Defense Front) in May 1994,[37] and of the *Red de Monitoreo Ambiental* (Environmental Monitoring Network) in September 1996. The *Red*, in particular, is a direct result of the awareness and information campaigns initiated by the members of the anti-oil network in the early 1990s (for instance, the 1995 Amazon Coalition—COICA workshop mentioned above), which picked up steam as the Texaco lawsuit evolved. International groups, however, were not the only ones promoting environmental training. Among indigenous peoples, for instance, monitoring of oil company activities and their impact on the local environment had been institutionalized by the CONFENIAE even before 1995. CONFENIAE's technical monitoring team continues to organize workshops among the grassroots with the goal of forming community teams for local environmental monitoring.[38] The establishment of the *Red de Monitoreo Ambiental* in the context of the Texaco campaign was important as evidence that environmental issues had become the unifying link among the diverse constituencies in the anti-oil network, particularly at the local level.

The *Red* was an initiative of *Acción Ecologica*, with the support of several international groups within the Amazon Coalition. *Acción Ecologica* trained community leaders to collect and assess data on the environmental degradation caused by Texaco operations. Environmental monitors also assessed the impact of one cleanup effort made by Texaco in 1995. The *Red's* monitoring effort was successful during the organization's first two years of existence, and it provided important information to plaintiffs' lawyers. The *Red*, however, has become significantly less active in the past few years. Staff directly involved in the monitoring effort admit that *Acción Ecologica*, after triggering the process, did not have either the institutional commitment or the resources to follow up on the initiative in the long term. The expectation was that the communities themselves would carry on the training of new monitors and maintain interest in the initiative, but it did not happen.[39]

The *Frente de Defensa de la Amazonia*, on the contrary, has increased and strengthened its role in the *Oriente* since 1994. The initiative is the result of work by activists that were directly linked to the preparation and filing of the Texaco lawsuit in New York. These *Oriente* activists were students in the United States at the time that attorney Bonifaz was gathering evidence for the case. They were eventually contacted by Bonifaz and signed on to the lawsuit. Upon their return to the *Oriente*, the activists formed the Committee of Plaintiffs against Texaco *(Comite de los Demandantes contra la Texaco)*, which was gradually expanded to include more than twenty-five *campesino* organizations of the northern *Oriente*. The *Frente's* work consists in lobbying authorities to cooperate with the lawsuit, disseminating information among interested parties, and maintaining a continuing flow of information between affected communities and their lawyers in the United States. The *Frente* has promoted conferences and courses on environmental and human rights, thus greatly contributing to a heightened awareness on these issues among *Oriente* communities.[40] Although indigenous organizations are not members of the *Frente*, they often support and join in its initiatives.[41]

While the creation of the *Frente* has benefited the anti-oil network's campaign against Texaco at several levels, it has also highlighted the persistence of old divisions within the network. Focusing on the positive, the *Frente*, with offices in towns in the *Oriente* (Sucumbios and Coca-Orellana), has been able to expedite communications with and between local groups in the Amazon. Because of the coordinating role of the *Frente*, it has become easier for organizations in Quito and abroad to remain attuned to the specific demands and nuances of expectations of local constituencies vis-à-vis the actions and goals of the network. On another level, the *Frente* identifies as one of its primary missions that of protecting the rights and the environment of Amazon people. Thus, it has contributed to alter traditional perceptions that settlers and *campesinos* are inherently in opposition to the goals of environmental preservation and are natural foes of indigenous peoples.

The *Frente* has faced two types of challenge that, to a certain extent, are the local level equivalents to challenges that have plagued the anti-oil network as a whole. First, the *Frente* has struggled to keep its constituency united in face of Texaco's pressures to negotiate a settlement. Second, it has had to assert the right of the plaintiffs to determine by themselves and without interference or pressure from any other actors—including their allies—the future of the Texaco lawsuit.

In 1995, Texaco and the government of Ecuador negotiated a deal by which the company committed to pursuing the environmental reparation of some of the areas affected by its operations. The agreement had the support of local authorities in the municipalities directly benefiting from the cleanup efforts and by some Quichua communities, who were offered (and eventually accepted) one million dollars in "compensation."[42] In return, the indigenous communities agreed to withdraw their participation in any political or legal action against the company. Texaco's money however, never benefited indigenous communities. It was squandered by the leadership, and some leaders eventually had to respond to charges of corruption.[43] It soon became evident that the deal was an attempt by Texaco to support its motion to dismiss the New York lawsuit on the grounds that it had become superfluous. The *Frente* was able to organize a general meeting in which the plaintiffs repudiated the Texaco's "settlement offer" and reiterated their commitment to pursuing the lawsuit.[44] Yet the possibility of independent deals with the oil company remains a potential threat to the unity of groups within the *Frente* and between the *Frente* and its indigenous allies.

The prospect of a settlement between the plaintiffs and Texaco has also become an object of contention between the *Frente* and both the Quito-based environmental organizations and the anti-oil network's international members. The *Frente* is committed to representing the interests of its constituency, whatever these interests may be. As the plaintiffs decided to pursue the Texaco lawsuit in 1995, they may decide to enter into an agreement with the company in the future. As Luis Yanza, the *Frente's* president, strongly asserts in a letter to *Acción Ecológica:*

> In effect, in the face of a possible proposal for dialogue from Texaco, we have initiated a process of information and reflection that will culminate in a workshop with community leaders, under the coordination of an independent international expert, with the goal of deepening our understanding of U.S. legislation. After receiving all this information, the leaders will be able to guide their bases along the path that they (the grassroots) find most convenient. (author's translation)[45]

Yanza also indicated his resentment at allegations (which he implied were made by environmentalists) that the *Frente's* leadership had opted for negotiations. In fact, while environmentalist organizations in Ecuador and abroad go through great pains to assert their support to the plaintiffs' decision-making autonomy, they also have specific interests vis-à-vis the Texaco campaign and lawsuit. The possibility of establishing a precedent for multinational corporations' environmental accountability would represent a victory not only for the anti-oil network in Ecuador but also for anti-oil activists throughout the world. A settlement would certainly jeopardize this prospect.

It is interesting to note that negotiations between affected communities and an oil company have been perceived differently in the context of the campaign against ARCO. Rather than indicating weakness, the capacity to negotiate with an oil giant was a sign of the relative strength of local groups. A brief discussion of the ARCO campaign should help clarify the reasons for these differences as well as their implications for the anti-oil network in Ecuador.

The Campaign against ARCO

The transnational campaign against ARCO's oil operations in Block 10, in the southwest part of the *Oriente*, the Pastaza region, was launched in December 1993. It happened during a meeting—the Villano Assembly—called by the Organization of Indigenous Peoples of Pastaza. While the transnational campaign against ARCO formally started in 1993, OPIP had opposed the company's operations since 1988.[46] Between 1991 and 1992, OPIP and ARCO engaged in negotiations that eventually proved to be unsatisfactory to the indigenous organization. One important outcome of these initial attempts to reach an agreement was the commissioning of an independent review of ARCO's ongoing and planned operations in Pastaza.

The process that led to the independent review reveals the breadth of the anti-oil network in Ecuador and its capacity to pursue several parallel, yet complementary strategies at once. On the one hand, the anti-oil network in the early 1990s was bracing itself for a major *confrontation* against Texaco. On the other, it was engaging in efforts to facilitate *dialogue* between oil companies and affected communities in Ecuador. In both cases, though, this comparative study of the campaigns reveals and reiterates the concern of international members of the anti-oil network that it remain accountable to the decisions of the network's national and local groups. In the case of the Texaco campaign, *Acción Ecológica* took the lead in devising a confrontational strategy to hold the company accountable for its actions in the *Oriente*. It received the unconditional support of the Amazon Coalition and others in the network. In the case of the ARCO campaign, when OPIP opted for negotiations, it also received the unconditional support of international groups, even from those that historically have opposed oil activities in the Amazon as a matter of principle, such as the Rainforest Action Network. In fact, a pool of resources that included RAN's and those from Oxfam America, OPIP, and CONAIE, as well as ARCO's made possible the independent review.[47]

The findings of the independent report, compiled by a team of experts from the University of California, Berkeley, criticized ARCO on several grounds.[48] The report's warnings, combined with the growing level of awareness that the Texaco campaign had generated among indigenous organizations in the south of the *Oriente*, contributed to the decision made at the Villano Assembly to avoid a repetition of the Texaco experience.[49] In the words of an OPIP leader, "What happened in the north is like a mirror for us . . . we cannot let the same thing happen here."[50]

Yet, rather than attempting to block oil exploration by ARCO, the campaign aimed at guaranteeing that indigenous peoples participated in environmental and social planning, in monitoring of company's activities, and in the economic benefits they generated.[51] For that to happen, company officials and indigenous representatives had to negotiate.

In 1993–1994, however, negotiations between ARCO and the indigenous communities of Pastaza did not fare much better than they had between 1991 and 1992. Problems derived from issues of legitimacy that originated both within and outside the anti-oil network. OPIP's legitimacy was undermined by its failure to follow up on initiatives to improve its technical capacity. If indigenous communities demanded participation in environmental planning, they needed to be able to present credible data and technically sound proposals.[52] To address these issues, OPIP created the Amazanga Institute with support from RAN. The goal of the institute was to build capacity among indigenous peoples on environmental matters by putting them in contact with researchers and scientists, and by providing state-of-the-art technology for environmental monitoring (such as Geographic Information System—GIS). The initiative was short-lived, however. The institute failed to accomplish its goals for several reasons, and its credibility was hurt by charges of misappropriation of external funding.[53]

OPIP's legitimacy was also threatened by ARCO's support of the Association for Indigenous Development (ASODIRA), which claimed to be the true representative of the indigenous peoples of Pastaza. To many, ARCO's dealings with ASODIRA were deliberate attempts to undermine OPIP through a "divide and conquer" tactic.[54]

Be that as it may, by early 1994 OPIP had unquestionably reestablished its position as the main representative of the indigenous peoples of Pastaza. With the help of domestic and international allies, the association continued to pressure ARCO to negotiate satisfactory conditions for its operations in Block 10.

Within Ecuador, OPIP counted on the support of *Acción Ecologica,* CONAIE, and CONFENAIE to lobby the Ecuadorian government. Ecuador's minister of Energy and Mines eventually came forward to mediate OPIP's negotiations with ARCO.[55] Internationally, Oxfam America, RAN, and the Seventh Generation Fund suggested and helped organize a meeting between OPIP leaders and ARCO officials at the company's headquarters in Texas.

By the end of 1994, negotiations between ARCO and OPIP had broadened in scope to include several other indigenous peoples organizations in Ecuador, such as CONAIE, CONFENAIE, and ASODIRA, Ecuador's Ministry of Energy and Mines, and Oxfam America (the latter as an observer). Two main proposals were agreed upon. First, indigenous organizations would develop a long-term regional development plan to be considered for funding by ARCO. Second, a technical environmental committee composed of Ecuadorian officials, ARCO personnel, and indigenous peoples' representatives would be established to determine the terms of an environmental impact assessment of ARCO's operations in Pastaza and the level of indigenous participation in that study.[56]

While these commitments were important as precedents for shaping ARCO's (and other oil companies') future behavior, they eventually became sources of frustration for local groups whose expectation had been heightened by the 1994 agreements. In fact, the agreements had few concrete consequences, for at least three main reasons. First, since 1995 ARCO has greatly reduced its investments in Block 10 (and in Ecuador in general). Its primary interest currently lies in an area called Block 24, which includes part of the Pastaza province and most of its neighboring province, Morona Santiago. Second, indigenous peoples' associations were unable to present a unified proposal for a regional development plan. ARCO has argued that until there is consensus among indigenous peoples it would not consider separate proposals. Finally, while the technical environmental committee issued a report in 1998, it failed to accomplish many of its initial objectives, at least concerning the expectations of indigenous peoples and their supporters. For instance, the committee sharply separated social and economic issues from environmental concerns. For some observers this decision undermined the very essence of the approach of local communities in the *Oriente* as well as that of OPIP, which asserts that "The Defense of Nature and Social Justice Are Inseparable" (OPIP's protest banner).[57] Despite their formal participation in the committee, indigenous peoples still felt that their role was limited by their lack of capacity and training to address complex technical issues. Last but not least, as of today, it remains to be seen how the committee's recommendation will translate into material actions.[58]

Despite frustrations with negotiations (or because of them), since 1998 the campaign against ARCO has gained momentum. More local organizations have joined and network members have reevaluated their strategies. While the precedent of negotiating with ARCO remains a valid option in the overall arsenal of strategies of the Ecuador's anti-oil network,[59] it has recently returned to confrontational initiatives against the company.[60] Three federations of indigenous peoples representing the Shuar and Achuar peoples in Block 24 have received the support of national and international groups to "unconditionally oppose oil development in their lands."[61]

One of the most interesting strategies implemented by one of the Shuar federations, FIPSE, was a legal action ("amparo petition") against ARCO.[62] The presentation of the demand in August 1999 was preceded by a march of indigenous peoples to the town of Macas, in protest against oil development in the region. In the legal action, FIPSE accused ARCO of violating the Shuar people's rights to organizational integrity. FIPSE denounced ARCO's attempt to divide the federation by negotiating independent deals with individuals and groups that are formally represented by FIPSE. In September 1999, the courts issued an order forbidding ARCO from approaching individuals or grassroots groups in the Shuar's territory without the formal authorization of FIPSE's general assembly.

The courts' decision in favor of the indigenous federation was an important symbolic victory for local groups and for the anti-oil network in general. Its merit is that it affirms that oil companies and governmental agencies can be made to respect the rights of local communities to self-representation and to a safe environment. While many such rights have been historically consecrated in the Ecuadorian Constitution, until recently they had been often ignored.[63]

Taking Stock of Ecuador's Anti-Oil Network

The example of the anti-oil transnational advocacy network in Ecuador supports existing arguments about the effectiveness of transnational advocacy networks. As discussed in the Introduction, transnational networks have been effective in changing the behavior of states[64] and international organizations.[65] In fact, the activism of the anti-oil network has affected the Ecuadorian government on at least three levels. Domestic and international lobbies have been able to guarantee that the government shall not oppose the continuation of the Texaco lawsuit in a foreign court. In addition, environmental activism has also succeeded in opening avenues of dialogue between Ecuador's Ministry of Energy and Mines and other agencies, and populations affected by oil operations. Finally, the network's legal strategies have slowly affected Ecuador's legal system and there has been a noticeable trend toward implementing existing environmental and human rights legislation that was frequently ignored in the past.

The network has also affected international organizations in both the public and private sectors. Public institutions, such as the World Bank, have increased their "sensitivity to the political consequences of involvement in the oil sector in Ecuador."[66] In the case of multinational oil corporations, some have become more aware of—and willing to respect—the social and environmental rights of affected populations. Activism has delayed activities, altered plans for pipelines, oil wells, and roads. It has also brought environmental impact assessments, once a mere formality, under closer social and governmental scrutiny. But the important question is: how effective has the anti-oil network been in protecting the *Oriente's* environment and its populations? Or, to put it differently, to what extent has local groups' participation in the transnational advocacy network affected their political and technical capacity to assert their environmental and social rights?

The campaigns against Texaco and ARCO, unfolding against the background of unprecedented indigenous activism in Ecuador in the 1990s, have amplified the political voice of indigenous federations and confederations in the country's policy decisions regarding oil exploitation in the *Oriente*. A more specific impact of the anti-oil network, however, has been the extraordinary increase among local populations—indigenous or not—of their level of environmental awareness and consciousness about their rights to a clean and healthy environment.[67] As a consequence, popular organization has intensified in the *Oriente* in the past fifteen years, an essential condition for political empowerment.

The formation of the *Frente para la Defensa de la Amazonia* is concrete evidence of this process. Not only does it legitimately represent the interests of a large number of *campesino* groups, giving them voice in national and international arenas, but it also constitutes a forum in which these groups have forged alliances (with indigenous peoples and environmental organizations) to pursue further struggles. In addition, the *Frente* has assumed the responsibility of continuing educational and awareness efforts on *campesinos'* rights. The *Frente* has played a crucial catalyst role for local civil society interests in the northern *Oriente*. Yet the very political visibility of the *Frente* has exposed it to two different sets of challenges. The first challenge is against its unity. Currently, the *Frente's* political strength greatly derives from its capacity to mediate between the interests of its environmentalist supporters (national and international environmental organizations), its indigenous allies, and the material expectations of most of its constituency, who anticipate some level of concrete compensation from Texaco. It is not clear whether or for how long these groups will remain bound by complementary goals vis-à-vis the Texaco lawsuit and related strategies. If and when such a degree of unity diminishes, the political future of the *Frente* and its role in contributing to environmental protection initiatives in the *Oriente* are uncertain.

Most troublesome, however, is the intensification of demands on the *Frente's* capacity as a result of the presence of an increasing number of oil companies in the *Oriente*. In the last decade, many larger oil companies once operating in the *Oriente* started to negotiate their concessions with smaller, less known companies (usually regional U.S. companies or oil corporations from Ecuador and other Latin American countries). There have also been increasing numbers of independent contractors in the *Oriente* working for larger companies. Ironically, this trend seems to be, at least in part, a reaction by large, highly visible, oil multinationals to increased social and environmental activism in the *Oriente*, which have raised the stakes of oil exploitation in the area. Monitoring the environmental standards of such a multitude of contractors and small companies has presented a challenge to the *Frente's* work that it is still not fully prepared to address.[68]

Technical capacity (or lack of it) seems to be the Achilles heel of local groups in the *Oriente*. The three attempts to promote or draw from local groups' technical capacity that have been discussed in this chapter—the *Red de Monitoreo Ambiental*, OPIP's Amazanga Institute, and the ARCO-sponsored environmental committee—have failed in the long term. This reveals, once again, the perverse dynamic that has often plagued transnational environmental advocacy networks. It is precisely due to their participation in the network's effort—and thanks to the success of these efforts—that local groups have gained a voice in decision-making processes of governmental agencies and private corporations, and a role as legitimate representatives of local civil society's interests. Yet the strengthening of local groups' political standing is not always accompanied by the technical capacity required for meaningful participation in these arenas. As a result, local groups may miss important opportunities to affect policy and sometimes even lose legitimacy in the process, which in turn inevitably compromises their political position.

Ecuadorian local groups have been able to compensate for their lack of technical capacity through the strength of their alliances with national members of the anti-oil network. Despite inevitable differences, local indigenous groups firmly rely on the political and material support of CONFENIAE and CONAIE for most of their initiatives. They have also profited from the legal support of organizations such as CORDAVI and the *Centro de Derechos Economicos y Sociales* (CDES)[69] to pursue lawsuits against oil companies and governmental agencies within Ecuador. National organizations, among them *Acción Ecologica*, have been instrumental in generating and disseminating important information on environmental issues among local groups, often with the help of its international partners, such as the Amazon Coalition and the Rainforest Action Network.

Finally, the impact of the anti-oil network on the *Oriente's* natural environment reflects the structural ambivalence that has characterized the network from its onset. There have been small gains toward environmental protection that can be linked to the actions of the anti-oil network and the Texaco campaign in particular. One example is the success of the Cofan indigenous group in protecting their lands and preventing oil operations by Petroecuador, which inherited Texaco's facilities after 1992.[70] In general, however, the *Oriente* remains open to oil exploitation and this situation will not change in the near future. While some members of the network would prefer to see the region closed to oil companies and propose ecotourism as its main economic vocation, others, particularly many local groups, do not oppose oil activities in principle. In fact, inside and outside the network there are those who argue that oil exploration may be an environmentally sound economic alternative for tropical forests. What must occur is that all necessary means be used to guarantee that exploration proceeds with minimum environmental disruption. While historically this has not been the case in Ecuador, OPIP's approach to negotiations with ARCO, for instance, indicates that this, rather than the end of oil activity, may be the goal of local groups. In other words, it would not be fair to assess the effectiveness of the anti-oil network in Ecuador on the basis of its ability to prevent the access of oil companies to the *Oriente*. This was never the overarching goal of the network for many reasons, among them the fact that there is no consensus on the issue among the network's local membership base.

Over time, what seems to have become a point of consensus among network members is that local populations in the *Oriente* have the right to participate in decisions, made by both the government and oil corporations, about the type of development promoted by these actors in the Amazon. Local groups have also determined the principles upon which their participatory demands are grounded. Development should be promoted in an environmentally sustainable way, by which the health and well-being of populations and the ecosystem are not threatened. In addition, local groups must receive a share of the economic benefits deriving from oil activities, either through the oil companies' direct funding of community development initiatives, or through the government's effective provision of social services. While these objectives are far from being achieved in the *Oriente*, the activism of the anti-oil transnational network has carved spaces for participation, created arenas for dialogue, raised awareness of environmental and social rights, and consolidated political alliances. These are important elements in the continuing struggle of *Oriente's* local groups.

7

Environmental Activism beyond Brazil II—The Struggle against Large Dams in India

Most people familiar with transnational environmental activism have either heard or read about the struggle against the Sardar Sarovar hydroelectric dam on India's Narmada River.[1] The struggle against the dam has unfolded at several levels, initially focusing on the rights of displaced people and subsequently challenging the very viability of the dam itself in environmental, social, and economic terms. Last but not least, the struggle against the Sardar Sarovar dam has challenged hydroelectric projects at the global level, questioning the rationality behind the erection of those structures that were once referred to as "monuments of Modern Civilization."[2]

In this chapter, and in accordance with the emphasis of this book, I analyze the activism of the Narmada transnational advocacy network in terms of its impact on local organizations and on the local political and natural environments. The case of the Narmada network is particularly interesting because local mobilization predates by many years the emergence of a transnational advocacy network. It thus provides an interesting comparison with cases such as both the Rondônia and the Ecuador's anti-oil networks, in which local mobilization occurred parallel to—and was often deeply influenced by—transnational activism. In addition, while the environmental preservation of the Narmada valley is an uncompromising goal of network members, social justice for affected populations remains their foremost priority. More than any other network discussed in this book, the Narmada network provides further insight on the role of local groups in framing and prioritizing issues and goals within transnational advocacy networks, and the relationships between these processes and different approaches to environmentally sustainable development (as discussed in the Introduction).

THE STRUGGLE AGAINST THE SARDAR SAROVAR DAM— BACKGROUND AND NETWORK ORIGINS

The Sardar Sarovar Project (SSP) is part of the largest river development scheme in the world. It was conceived within a plan to build more than three thousand small, medium, and large dams along the Narmada River valley. The Sardar Sarovar Project, which includes a dam and power generation facilities, a canal, and an irrigation network, affects the Indian states of Gujarat (where the dam and irrigation canal are being built), Madhya Pradesh, and Maharashtra (both states are home to most of the population affected or to be affected by the creation of the reservoir). One of the most controversial aspects of the Sardar Sarovar Project is the number of people that it will affect.[3] Estimates range from 100,000 people in 245 villages[4] to 200,000[5] and 325,000,[6] to mention just a few numbers. Opposition to the dam dates back to the 1960s, but then it was mostly confined to technical debates among the three states involved.

Popular resistance against the Sardar Sarovar dam picked up steam in the late 1970s. It is interesting to notice that it originated among different sectors of the local population, whose motives for resisting the dam also varied. Naturally, residents of the villages to be flooded by the reservoir questioned the dam. Some were primarily concerned with the project's resettlement and rehabilitation (R&R) provisions and with the conditions for relocation. Others resisted relocation in principle and were not willing to abandon their homes regardless of the terms of the R&R plans. Different groups supported the resistance efforts of these populations.

The NGO Arch-Vahini, for instance, was actively involved in supporting villagers in the state of Gujarat who faced submergence. This group's primary demands were for an improved resettlement and rehabilitation plan and equal rights of R&R for grown male children and for those who did not have formal title to their lands.[7] In Madhya Pradesh, many Adivasi[8] villages had come together in the early 1980s through the activism of the Sangath movement. The movement aimed at securing Adivasis' access to public forests and land, on which their very subsistence depended. The Sangath eventually realized that many of its member villages would be submerged by the Sardar Sarovar reservoir. It framed its mobilization efforts mostly in terms of the dam's ecologically unsustainable and socially unjust impacts.[9] Also in Madhya Pradesh, a well-off group of farmers and traders, or Patidars (upper caste), from the

Nimar valley, came together under an organization called Nimar Bachao Samiti. Finally, in the state of Maharashtra, affected people were familiar with the workers' union Shramik Sanghatan, which provided a starting point for organization against the dam.[10]

In 1985, Medha Patkar, a social worker and activist, started to work with popular organizations in Madhya Pradesh and Maharashtra toward strengthening their resistance initiatives against the Sardar Sarovar dam. Arch-Vahini and most groups of Gujarat's affected populations did not participate in this effort. From the start their concern was with reforming R&R policies rather than preventing the construction of the dam.[11] Patkar's work counted on the support of a large number of regional and national activist and advocacy organizations in India, among them the Lokayan network, a forum for interaction between activists and concerned intellectuals, the SETU, a network of socialist activists in Maharashtra, and the Tata Institute of Social Science.

In 1987, local resistance against the Sardar Sarovar Project attracted the interest of the international environmental NGOs at the forefront of the Multilateral Development Banks campaign. In September of that year, Medha Patkar traveled to Washington, D.C., at the invitation of campaign activist organizations. On her agenda were discussions with World Bank executive directors about the role of the bank as the main external financing institution for the SSP. As a result of her trip, Patkar became aware of the "substantial possibilities of using the Bank as an arena and even instrument of campaigning," and of the clout of environmental groups in the North.[12] Around the same time, grassroots groups in Madhya Pradesh and Maharashtra, under the leadership of Patkar and other activists, were forming an umbrella organization to coordinate their resistance, the Narmada Bachao Andolan (NBA). These two events in combination marked the emergence of the Narmada transnational advocacy network.

To what extent, however, was the Narmada network an *environmental* advocacy network? There have been questions raised about the accuracy of characterizing the resistance against the Sardar Sarovar dam as motivated by environmental considerations.[13] In fact, the predominant focus of the struggle until 1988 was on resettlement policies and reparations for affected populations, rather than on an explicit opposition to the project and its environmental consequences. In 1988, however, the Narmada Bachao Andolan and its affiliated organizations declared their total rejection of the project, as conveyed by the slogan, "No one will move! The dam will not be built!" The justification for such a radical rejection was that the project was environmentally and economically disastrous.[14] In 1989, the NBA organized the National Rally Against Destructive Development, which became known as "the coming of age of the Indian environmental movement."[15] In it, the NBA defined its fight as one to "end all projects which devastate the environment and destroy people's livelihoods . . ."[16]

Further justifying the study of the Narmada network as a transnational environmental advocacy network is the fact that the main international supporters of the Narmada struggle were environmental NGOs such as the Environmental Defense Fund, the National Wildlife Federation, and the Environmental Policy Institute, the leading organizations behind the MDB campaign. Friends of the Earth and the International Rivers Network (IRN) also became key international players in the late 1980s. Given the extent of the environmental impact of the Sardar Sarovar Project, it is not surprising that the struggle against it has acquired environmental connotations. Concerns include the threat of diseases, damage to fisheries, deforestation, loss of biodiversity, soil erosion, and waterlogging and salinity in the reservoir area.[17] A major contention regarding the SSP is whether the project is conceptually flawed in environmental terms, or whether its environmental impacts can be mitigated by the adoption of appropriate measures.

Under the skillful coordination of the Narmada Bachao Andolan, the resistance against the Sardar Sarovar Project gained further national and international visibility. International recognition in the form of environmental awards to the NBA, such as the Right Livelihood (1991) and Goldman (1992) awards, also helped in this process. One consequence of such an increased visibility was that a growing number of individuals and organizations lent support to the movement. Among the members of the Narmada network in the early 1990s were selected World Bank staff, foreign and Indian intellectuals, scientists, artists,[18] and renowned Indian activists. Among the latter is Baba Amte, one of India's most respected moral leaders. His participation has highlighted the nonviolent nature of the resistance against the SSP and its commitment to Gandhian principles. In addition to individual activists, the Narmada network counts among its members sectors of the international and Indian media, selected Indian politicians and parties of the Left, and most recently, sectors of the nonresident Indian community around the world, particularly in the United States.[19]

Given the composition of the Narmada network, it is not difficult to conclude that it has been endowed, almost from its inception, with significant political, material, and technical resources. In comparison with both the Rondônia and Ecuador anti-oil networks, it has been, by far, the most resourceful. Its legitimacy has gone unquestioned since the network evolved around a preexisting set of local groups who have resisted the SSP almost from its inception. The nature of the Narmada Bachao Andolan as an umbrella coalition of grassroots groups has institutionalized local participation in the network. The fact that the NBA counts on the financial resources provided by the Patidars, who represent a significant portion of its affiliates, has preserved its autonomy vis-à-vis other network members, particularly its international supporters. This financial autonomy, together with NBA's position as the catalyst for most of the network's initiatives at the local and national levels, have contributed to the network's capacity to avoid charges that often plague transnational networks (for instance, that local groups are being manipulated by foreign interests).

The inclusion of the Sardar Sarovar Project among the case studies of the MDB campaign granted to the Narmada network a whole array of political resources, such as a larger network of international allies, international visibility, and access to international arenas, such as the World Bank board of directors, U.S. and European legislatures, and Japanese financial institutions.[20] Finally, the participation of national research and advocacy think tanks, such as Lokayan and the Tata Institute, in the network made available to it the technical expertise of academics and scientists. These national organizations also helped to "nationalize" the Narmada struggle, framing it in terms that would resonate within larger sectors of the Indian public opinion.

Network Cleavages or Concurrent Goals?

The very dimension of the Narmada network in terms of the numbers of people directly and indirectly affected by the Sardar Sarovar Project, the groups that represent them, their national and international allies, and concerned activists, suggests that the network has had to contend with an unusually high volume of issues and interests within it. Evidently, there were and have been divisions among network members. Yet a further complicating aspect in the analysis of the Narmada network is to distinguish where the actual divisions lie and where differences in intensity of preferences and focus of interest merely disguise concurrent or complementary goals. For the sake of clarity, rather than discussing every disagreement that ever occurred among the members of the Narmada network, I will focus the analysis on three issues: class differences and the legitimacy of activists, the internationalization of the struggle, and emphasis on environmental versus resettlement issues.

A network as vast and inclusive as the Narmada congregates individuals from all social classes. Class differences affect the network on at least two levels. Locally, that is, among the populations directly affected by the Sardar Sarovar Project, one finds both Adivasis and Patidars. The everyday relations between these two groups are characterized by some analysts as consisting of class tensions, where Patidar landowners exploit Adivasi labor.[21] Different from many social movements in India, the Narmada movement has been concerned with addressing these internal differences. The struggle waged commonly through the NBA against losing their land, life-styles, and natural resources such as fertile soils and forests, has helped these groups to temporarily transcend their differences. Problems, however, still remain[22] and it is not entirely clear how they affect the internal dynamics of the NBA and its affiliate organizations. I suggest that class differences affect the movement's decision-making hierarchy, overlapping with the natural segmentation between leaders and the rank and file.

A different type of class division (one not limited to Adivasis and Patidars) also affects the Narmada network's choice of priorities. For local groups it is evident that their struggle is about avoiding the loss of their land, communities, and lifestyles. For middle-class activists and intellectuals (residing or not in the Nimar valley) the Narmada struggle has both a local and a national dimension, but their focus tends to be on the latter. For them, the struggle is an opportunity to challenge India's environmentally and economically unsustainable, and politically and socially undemocratic, development model. This clear difference within the network has made it a target of critics,[23] who perceive the Narmada struggle as one manipulated by India's left-wing intelligentsia.

The internationalization of the struggle against the Sardar Saravar dam was also a controversial topic within the Narmada network.[24] There was divergence on the merits of discussing national problems in international arenas. Those who opposed a campaign for World Bank accountability vis-à-vis the SSP feared that it would compromise efforts to establish a constructive engagement with the Indian government. On at least one occasion, for instance, the Narmada Bachao Andolan was criticized for lobbying the World Bank to prevent the Indian government from making forest land in Maharashtra available for resettlement. While this strategy made sense given the environmental concerns of MDB campaign's and NBA's activists, for local populations willing to relocate to the Taloda forest it represented a lost opportunity.[25]

The focus of the Narmada struggle on the World Bank also posed strategic demands on the NBA that challenged its relationship with the grassroots. The campaign against the bank forced the NBA to reformulate its stated objectives periodically in order to make the best possible use of available opportunities. Explaining successive changes in campaign objectives to local populations was sometimes challenging. During the seven years of the Narmada international campaign (1987–1993), demands varied from calls for better resettlement and rehabilitation provisions, to the interruption of the project and withdrawal of World Bank financing, to requests for an independent review of the project and of the bank's role in it. The latter was, in principle, an opportunity to establish a common baseline for dialogue among interested parties on alternatives to improve the project's environmental and social performance.

Finally, the internationalization of the Narmada campaign generated controversy inside and outside the network about its hierarchy of priorities. No one denies that mobilization in the Nimar valley began as a result of concerns about the fate of affected people: could the flooding of the valley be avoided? Could resettlement be avoided? If not, what were the rights of relocation and rehabilitation of affected populations? The process of internationalizing the struggle added another layer to these initial concerns—that of issues related to the environmental preservation of the valley and to India's development model. Evidently, these issues do not contradict, but rather complement each other. It is important to notice, however, that the sheer magnitude of the task of addressing these problems forced a certain division of labor both within the network and within the NBA itself.

On the one hand, local populations and grassroots groups within the NBA continued to articulate campaign objectives and initiatives in terms of affected peoples' rights to their lands and lifestyles. On the other, NBA leaders closer to international organizations, India's national activist networks, and international environmental groups focused their critique of the SSP on the structural issues that it raised about India's and the World Bank's commitment to environmentally sustainable development and democratic processes. It is not difficult to understand that this division of labor could, on occasion, be perceived both internally and externally, as a local/international cleavage or a distancing between network leadership and grassroots.

The Narmada Network Strategies and their Local Impacts

One of the most interesting aspects of the Narmada network was the capacity of its members, but particularly that of the Narmada Bachao Andolan, to pursue simultaneous, yet completely different strategies at the local, national, and international levels. While this is an achievement on its own merits, I argue that, in retrospect, these strategies have been more successful in challenging India's and the World Bank's development model than they have been in guaranteeing the immediate interests of affected populations in the Nimar valley. In the paragraphs below I discuss examples of strategies implemented at each level and their impacts on the local people.

The NBA's local strategies aimed at increasing the power of resistance of both individuals and local populations as a whole. Disseminating information about the Sardar Sarovar Project and raising awareness about villagers' rights has fortified individuals' resolves: many have vowed to remain in their houses despite threats of rising waters; others have gone into hunger strikes until the government answers to their specific demands; still others have committed to sacrificing their lives by jumping in and drowning in the Narmada waters if and when the flooding of the valley becomes inevitable.[26] The NBA's efforts to organize and raise villagers' awareness led to a series of mass demonstrations throughout the 1990s, and the most recent *satyagraha,* a non-violent rally by affected people, occurred from July to September, 2000.

The NBA organized the movement's most famous *satyagraha* to date in December 1990. Ten thousand villagers and activists walked to the site of the dam willing to stop works through nonviolent means. At the border with the state of Gurajat, they were met by a police blockade and a counterdemonstration organized by that state's government. The villagers camped by the roadside and Patkar and six other villagers and activists initiated a hunger strike.

Activists, Patkar in particular, and villagers have used hunger strikes on other occasions as instruments of leverage. Other strategies devised and implemented by local people are the selection of individuals to remain in areas designated for flooding at specific points in time ("Save or Drown Squads"), and the nonviolent resistance to police repression.[27] These strategies, with their dramatic content and unquestionable indication of how much is at stake for local populations, have lent both legitimacy and visibility to the efforts of the Narmada network. Yet they also have generated fierce governmental repression in the valley.[28] The most common forms of violence against villagers are arrests on false charges, beatings, and forced eviction from villages in the flooding zone.

The one concrete "positive" impact of the individual sacrifices and risks that were borne by villagers throughout the 1990s was that they captured the world's attention. Local activism greatly contributed to the leverage that the Narmada network imposed on international financing organizations funding the SSP. For instance, the Japanese government withdrew its support to the project in 1990, claiming public opposition as the justification for its decision.[29] The campaign against the participation of the World Bank in the SSP was longer and with consequences that went beyond the Nimar valley and India itself.

The international mobilization against the bank started between 1987 and 1988, coinciding with Metha Patkar's first visit to Washington, D.C. The Environmental Defense Fund took the lead in organizing pressures against the World Bank's role in the SSP, and eventually organized a network of activist groups in North America, Europe, Japan, and Australia, known as the Narmada Action Committee. In 1989, the committee's lobby in the U.S. Congress arranged for Patkar to address congressional representatives in a hearing about the SSP. Patkar succeeded in sensitizing the audience, and many congressmen wrote to the World Bank demanding the suspension of its participation in the SSP.[30]

In an interesting indication of its political knowledge of and sensitivity to the Indian political context, the NBA slightly changed the tone of its demands in the early 1990s. Lobbying for the cancellation of the SSP and for the World Bank's withdrawal generated significant opposition against the NBA within India. Campaigners were accused of halting the march of progress and preventing drought-prone areas in Gujarat from having access to the Narmada waters. During the "long march" of December 1990, the NBA decided to abandon a confrontational posture against the government and adopt, instead, a more "neutral" position of demanding an independent review of the SSP. As Patkar explained, this would provide a "way out" for the government, who could, in theory, respond to popular concerns voiced through and mediated by neutral moderators working within a commonly accepted framework.[31] In response to Patkar's and other activists' twenty-six-day hunger strike following the long march, and to the intense lobbying of international activist organizations, the Indian government promised an independent review (which was later called off) and the World Bank followed suit.

The independent review of SSP commissioned by the World Bank was performed by a team of experts under the leadership of Bradford Morse, former director of the United Nations Development Program (UNDP). The panel's conclusions, known as the Morse Report, became available to the bank and interested parties in June 1992. They clearly vindicated the NBA's criticism to the Sardar Sarovar Project declaring that the project was "flawed, that resettlement and rehabilitation of all those displaced by the projects is not possible under prevailing circumstances and that the environmental impacts . . . have not been properly considered or adequately addressed."[32] The report urged the bank to withdraw from the project. Rather than accepting the panel's recommendation, the bank issued a document in which it listed a series of remedial measures for the problems of the SSP identified by the Morse Report. It also initiated negotiations with the Indian government for the implementation of such measures. Neither the NBA not its international supporters accepted such responses, essentially doubting that remedial measures would address the structural flaws of the SSP. Instead, they continued lobbying the bank to withdraw from the project. As a direct result of the Narmada network's efforts,[33] and of further evidence that the bank's remedial measures had been ineffective, the bank withdrew its participation from the project in March 1993.

The World Bank's abandonment of the Sardar Sarovar Project represented an enormous victory for the Narmada network. It also made evident the power that local organizations such as the Narmada Bachao Andolan could exercise at the international level with the support of international allies. The World Bank's withdrawal, however, was a political victory. It was not a conquest that benefited populations in the Nimar valley in the long term. Lack of World Bank funding did delay the construction of the dam, but eventually the Guajarat state and the company in charge of construction overcame their liquidity problems. During the 1993 and 1994 monsoons several villages were submerged or partially submerged. In fact, increased repression in the valley seems to have been a response by regional governments to the Narmada movement's international victory,[34] as summarized below.

State-sponsored repression against activists and villagers has been a permanent aspect of the struggle against the social and environmental consequences of the SSP. Its intensity, however, has varied across time and space. In the state of Gujarat, for instance, repression has been historically more limited than in the states of Maharashtra and Madhya Pradesh. Gujarat, it should be highlighted once again, will benefit most from the hydroelectric scheme at several levels: the SSP promises to bring water to the arid regions of the state (although critics say that the poor will have no access to it); large landowners have cut handsome deals when selling property to be used for resettlement purposes; and infrastructure linked to the SSP, such as roads and canal works, have created economic opportunities for local businesses and jobs for local citizens (although in a lesser extent than was promised to resettled individuals). Not surprisingly, Gujarat administrations have faced significantly less resistance and social unrest against the SSP than the governments of the other two states.

Most of the villages to be flooded by the SSP reservoir are within the borders of Maharashtra and Madhya Pradesh. Authorities in both states have used the police and other organs of the repressive apparatus to curb mobilization against the project. The police have intimidated activists and villagers, sometimes even resorting to torture and illegal detention, and created obstacles to the organization of meetings, *satyagrahas*, and other forms of popular protest.

In India, the police—under the jurisdiction of the various state governments—have a low level of professionalism (particularly when compared to the military), and have been used frequently to support the interests of dominant groups. The area affected by the SSP is one of the major strongholds of the Bharatiya Janata Party (BJP), which has been a major advocate of the dam. The steady increase in the levels of violence against opponents of the SSP in the 1990s coincided with the BJP's growing influence in national politics.[35] Efforts to mute popular criticism were supported by influential party members in New Delhi, who also expected to gain politically from portraying the SSP as an instrument of social and economic modernization. Finally, the nationalist rhetoric of the BJP provided strong ideological justification for local repression against the anti-dam movement, particularly in the early 1990s, when international activism against the SSP climaxed, the World Bank abandoned the project, and the Narmada scheme became a sensitive issue in India's foreign affairs.

While the Narmada network's strategy of lobbying the World Bank to withdraw from SSP did not improve the conditions of local populations, it did have profound impacts at national and international levels. In part as a consequence of the Narmada campaign, the World Bank initiated in-depth reviews of its policies on large projects.[36] The establishment of the inspection panel, for instance, was one outcome of this internal process as well as of continuing demands from environmental and human rights activists for the institutionalization of accountability and transparency procedures.[37] Finally, in 1997, the World Bank participated in launching the World Commission on Dams. The commission was set up as a response to concerns of civil society organizations, governments, multilateral institutions, and the private sector, with the "growing polarization between proponents and opponents of large dams (that) resulted in a virtual breakdown of constructive dialogue." Work by the twelve commission members, a group that includes Metha Patkar, "centered on reviewing the effectiveness of large dams and developing standards, criteria, and guidelines to advise future decisionmaking."[38] The commission issued its final report, "Dams and Development," in November 2000.

There is a striking difference between the achievements of the Narmada network at the local and at the international levels. It is almost as though one is observing the evolution of two parallel, yet unrelated, sets of initiatives that happen to refer to a single region of the globe. I will return to this discussion in the last section of this chapter, after looking at the strategies of the Narmada network vis-à-vis the Indian government and India's national development policies.

The Narmada Bachao Andolan never abandoned the demand for an independent review of the Sardar Sarovar Project that it had made to India's government during the 1990 long march. Eventually, in the wake of the World Bank's withdrawal from the SSP, the Indian government commissioned a five-member group to review the project. The group heard the opinions of more than one hundred experts on several issues related to the SSP. This process alone raised awareness among India's public opinion to the controversy surrounding the project. The five-member group issued its final report in 1994, but the government of India forbade its public release.[39]

Encouraged by the conclusions both of the Morse Report and of the Five-Member Group Report, the NBA filled a writ petition with India's Supreme Court in May 1994. Throughout 1995, the court held hearings, studied documents containing allegations that the SSP violated several rights of local villagers, and issued orders on the case. The legal history of the involvement of the Indian Supreme Court in the struggle against the SSP is beyond the scope of this chapter. Suffice is to say that the court battle became the focus of the NBA's struggle following the campaign for the World Bank's withdrawal from the project.

The NBA's court strategy had some concrete, albeit short-term positive results for the people in the Nimar valley. One of the supreme court's first decisions, in January 1995, was to mandate the suspension of works at the dam site until there was progress on resettlement and rehabilitation initiatives.

The order had a direct effect in preventing an increase in the number of villages submerged during the monsoons. It did not alter, however, the plight of populations that continued to be displaced by infrastructure work related to the dam, in particular the digging of the SSP main canal and irrigation network in Gujarat.

As the legal debate on resettlement issues continued, the state of Madhya Pradesh made an unprecedented request for the supreme court to mandate a ceiling of 81.5 meters as the maximum height of the dam wall (its actual height in 1995). The reason for such a request was the state's acknowledgment that it had neither the resources nor the land available to resettle the number of people that would be displaced by a larger reservoir (one created by a higher wall). In May 1995, the court granted Madhya Pradesh's request in an interim order and, until recently, had sustained its decision despite applications by the Guajarat state urging it to vacate that stay.[40]

These extraordinary legal victories of the Namada movement were reversed on October 18, 2000. India's supreme court delivered its final ruling allowing immediate construction on the dam up to a height of 90 meters. It also allowed the wall to go up to its planned height of 138 meters, pending periodical approvals by a state committee on relief and rehabilitation (Relief and Rehabilitation Subgroup of the Narmada Control Authority).[41] Since the ruling, work on the dam has gone forward. Further flooding of villages in Madhya Pradesh and Maharshtra did not occur only thanks to the severe drought that has affected the region in the last three years. The institutional context in which the final decision on the NBA's writ petition was made is relevant for an appreciation of the challenges that the Narmada network faces at the local level.

India's parliament and supreme court have had a difficult coexistence since the promulgation of the country's constitution, shortly after its independence from Britain. At the core of such difficulties is the contradiction between the principles of parliament's sovereignty and judicial review.[42] The history of the NBA's writ petition to the supreme court illustrates the tensions between the two institutions. By determination of the parliament, the Narmada Waters Dispute Tribunal was established in 1969 to settle differences and mediate the distribution of costs and benefits between the states affected by the SSP. The tribunal eventually produced an award, which determined a series of obligations, among them, those related to resettlement and rehabilitation. The NBA's lawsuit, which in 1995 had the support of the government of Madhya Pradesh, charged the central government and the state of Gujarat with proceeding with the SSP work while neglecting their R&R commitments. When in December 1995 and again in March 1997 India's supreme court ruled against the latter, ordering the interruption of construction works in the SSP dam, lawmakers were furious.[43]

Yet, this apparent autonomy of the supreme court vis-à-vis the parliament was short-lived and may have been a consequence of a period of realignment of political forces at both national and sub-national levels. After dominating India's politics from independence to 1989, a fractured Congress Party was back in control of India's central government in 1991. In 1995, however, when the first ruling of the supreme court interrupted the SSP, the party was about to loose its parliamentary majority once again. Similar instability affected the BJP during this period. The party, dominant in the states affected by the SSP between 1991 and 1993, lost power to the Congress Party in Madhya Pradesh, in November 1993.

The BJP's successful electoral campaign in the states of Madhya Pradesh and Gujarat (among others) in 1996, and at the national level in 1998, established a new balance of political forces in India. This relative stability helped reinstate a trend, inaugurated during the Indira Gandhi years, of limiting the court's independence. The BJP strength at regional and national levels is likely to have influenced the unanticipated final decision of the supreme court in 2000 against the NBA and in favor of the central government and of the state of Gujarat.

A final strategy pursued by the NBA and local supporters of the struggle against damming the Narmada is to broaden the debate within the Indian political and social establishment beyond the Sardar Sarovar Project. This has been done not only through mass mobilization but also by continued work with media channels to disseminate information about the struggle nationally. The process has counted on the support of many Indian intellectuals and artists. It has made its way to the Indian parliament, where political leaders have debated the merits of the supreme court's intervention in the SSP controversy.[44] India's overreliance on large dams has also been questioned by leftist parties,[45] although some activists resent these parties' low level of involvement with the issue, and with the SSP struggle in particular.[46] Despite its firm opposition to India's official policy on dams, the NBA, together with other groups working with displaced people in Madhya Pradesh has, since 1995, engaged in dialogue with the state government. This dialogue has centered on the formulation of a just resettlement policy as well as on a shift in the development process toward lower levels of social and environmental disruption.[47]

It seems evident that at both the national and international levels the Narmada network has evolved from questioning the social and environmental viability of the SSP to formulating a powerful critique of undemocratic, top-down, statist development.[48] In the words of Roy:

> In India over the last ten years the fight against the Sardar Sarovar dam has come to represent far more than the fight for one river. This has been its strength as well as its weakness. Some years ago, it became a debate that

captured the popular imagination. That's what raised the stakes and changed the complexion of the battle. From being a fight over the fate of a river valley it began to raise doubts about an entire political system. What is at issue now is the very nature of our democracy.[49]

Despite the merits of such an option, however, it is not clear that it has brought immediate gains to local populations. It is Roy, again, who best summarizes the impact of the Narmada network's emphasis on structural development issues beyond the SSP:

For the people of the valley, the fact that the stakes were raised to this degree has meant that their most effective weapon—specific facts about specific issues in this specific valley—has been blunted by the debate on the big issues.

Other analysts have agreed with this assessment, even when they have praised the Narmada Bachao Andolan and its allies' strategic choice of criticizing mainstream politics in India and of offering a program of action aimed at reform.[50]

Taking Stock of the Narmada Network

"The struggle in the valley is tiring...."[51] How is it possible that the Narmada network, arguably the largest in terms of number of members at local, national, and international levels, the most internationally visible, and the most resourceful of the three networks discussed in this book, seems to have produced the least significant local impacts? The issue is all the more striking when one acknowledges that the Narmada is, possibly, the most effective transnational advocacy network (among the ones studied in this book) in its impact on the behavior of nation-states, multilateral organizations, and the global civil society.[52]

The Narmada network was extremely successful in changing the behavior of the World Bank. It contributed to unprecedented decisions, such as the commissioning of an independent review of the SSP, the bank's eventual withdrawal from the project, and an entire reassessment of the institution's lending policies, culminating with its creation of the inspection panel and its active participation in the organization of the World Commission on Dams. The network also influenced the behavior of the Japanese government and of its international lending agency, the OECF. There are also signs that the network has affected perceptions at all levels of the Indian government. At a minimum, it has brought the voices of often ignored and "invisible" sectors of India's population, such as the Adivasis, to the country's highest judicial body. It has influenced the decision of the Madhya Pradesh state to oppose increasing the height of the dam wall. Such a decision was based entirely on that state's recognition of the resettlement challenges it would have to face in the event of a larger reservoir. The Indian state has even begun to discuss a national law on resettlement and rehabilitation issues when no such initiative existed before. The Narmada network has succeeded in dividing the government establishment regarding the role of large dams and of the Narmada River scheme in India's development. The Ministry of the Environment, for instance, and several legislative representatives at state and federal levels, have opposed the SSP. Finally, the Narmada struggle has played a crucial role in strengthening movements against large dams throughout the world.

Yet the question remains: To what extent did the actions of the Narmada network succeed in guaranteeing the social and environmental rights of its local membership base? To what extent was it effective in preserving the Nimar valley's environmental integrity? The answer, unfortunately, is "not much." Despite the interruption of work between 1993 and 2000, dam construction is currently proceeding at full force. Related project infrastructure never stopped and has continued to cause the displacement of thousands of individuals. Resettlement and rehabilitation provisions, even the relatively progressive ones agreed upon by the state of Gujarat at the onset of the SSP, never materialized. As a result, a growing number of villagers who had once accepted relocation have returned to their previous homes and villages in the Nimar valley. The monsoons have continued to submerge houses, belongings, and crops of people living at the banks of the Narmada. State-sponsored repression has continued, and sometimes intensified, through the years.

More important than asking whether the local objectives of the Narmada network were accomplished is to ask why they have not, particularly in light of the network's extraordinary accomplishments at the international and national levels. This question is important because it relates to one of the core arguments of this book, namely, that the effectiveness of a transnational environmental advocacy network is a function of the level of participation and empowerment of its local membership base.

A superficial analysis could make the case that the Narmada network invalidates this argument, given the undeniable political and technical strength of the Narmada Bachao Andolan and its catalyst role in defining the network's objectives and strategies. The challenge to the argument would sound somewhat like this: "The Narmada network's local membership base, i.e., the Andolan, was a powerful network actor from the onset of mobilization, it was further empowered by the national and transnational alliances that were formed around the opposition against the SSP. Yet, the Andolan, and the Narmada network as a whole, remained incapable of accomplishing their local goals." If this challenge were to be accepted, the claim that a network's effectiveness depends on the strength and level of influence of its local membership base would have to be abandoned, and explanations for the limited impact of the Narmada network at the local level would have to be found elsewhere.

The response to this challenge, and the chance of salvaging the argument, rests on a deeper analysis of the nature of the Narmada Bachao Andolan and of its role throughout the evolution of the Narmada network. A comparison between the NBA and the Rondônia Forum may clarify some important issues. The history of the Rondônia Forum could be summarized as that of a clearinghouse organization, created somewhat artificially to provide coordination and resources to grassroots groups that were at their infancy at the onset of the Rondônia network. As the network evolved, forum members increased their mobilization capacity—an asset obtained, mainly, as a result of their participation in the forum and in the network. Grassroots empowerment reached levels that even allowed these groups to challenge the legitimacy of the forum's leadership to "speak for them" (at one point in the evolution of the Rondônia network, grassroots organizations decided to speak with their own voices, without mediators, including the forum itself). One of the consequences of the eventual resolution of the forum's legitimacy crisis was that its role as a catalyst of local initiatives decreased, and that space was filled by specific grassroots organizations, such as those representing Rondonian rubber tappers and rural workers. The forum's diminishing role in the Rondônia network was at once a cause and a consequence of grassroots groups' increased assertiveness in the local context. Among the consequences of this increased assertiveness, or of their level of empowerment, was the increased ability of local grassroots groups to formulate objectives that prioritized local interests, transforming such objectives into the paramount goals of the network as a whole. Through the network's activism local groups eventually accomplished, or partially accomplished, their objectives (the specific example here being the reformulation of the Planafloro project and the volume of resources directed to the Program for the Support of Community Initiatives, PAIC).

The process with the Narmada Bachao Andolan seems to have unfolded in the exact opposite direction. While the evolution of the Rondônia network led to the eventual weakening of the forum's role and a simultaneous strengthening of its member organizations, the evolution of the Narmada network generated an unprecedented strengthening of the NBA, but such a process did not "trickle down" to the Andolan's constituent organizations. One of the consequences of this stalled process was the absence of venues for local grassroots groups to assert their priorities within the network and to influence other network members' approaches to environmentally sustainable development and pragmatic ways of making the concept an operational reality in the context of the Nimar valley. For instance, if dam construction and displacement are inevitable, what instruments can the network use to guarantee the improvement of the quality of life of affected individuals? While the NBA has directed resources toward this area, it has not been the center of its efforts.

The key question that the NBA's undeniable process of empowerment raises is one about the nature of the organization itself. Is it an organization that supports the struggle of grassroots groups in the Nimar valley or is it a representative of such groups, that is, a grassroots organization that speaks for the affected people and has a mandate to represent their interests? Once again, the comparison with the Rondônia Forum is relevant. Forum leaders never claimed that the organization was, nor was it ever perceived as, "representing" local grassroots groups. It has been, from its inception, an organization that claimed, on occasion, to speak "on behalf" of sectors of Rondonian civil society. Most frequently, however, the forum characterized itself as an umbrella organization or clearinghouse that facilitated alliances and common initiatives among different groups in the Rondonian civil society. The nature of the NBA, at least for outside observers, is not as clear. Some firmly assert that the organization is not one working "on behalf of the local people, [but one that is] primarily a body of the affected people, with some—very few—activists from outside—these too living in the valley itself among the people...."[53] If this is in fact the case, the puzzle of the reasons for the Narmada network's limited local effectiveness remains.

Yet, if one characterizes the NBA as a coordinating umbrella organization that supports local groups (village committees, unions, environmental and human rights groups, among others) then, while the NBA as an organization may have become further empowered as a result of its participation in the Narmada network, the same is not necessarily true of its affiliated organizations. The characterization of the NBA as an umbrella organization or clearinghouse seems to be prevalent in the literature. Baviskar defines it as a federation of "mass-based organizations from all over India." He explains that the "Andolan is at the forefront of such a coalition . . . of organizations that have formally recognized their common agenda, and the importance of coordinated action."[54] Fisher also defines the NBA as a "national coalition of environmental and human rights activist groups, scientists, academics, and project-affected people who espouse Gandhian nonviolent resistance techniques and stress their grassroots membership and goals."[55] Even Medha Patkar draws a distinction between the NBA as an organization of activists, and the specific institutions of representation of local populations.[56]

If a distinction can be made between the identity of the NBA and those of its member organizations, then the puzzle of the Narmada network's limited local effectiveness is solved. Throughout this chapter it became evident that as the NBA deepened its involvement in the Narmada network, becoming increasingly visible and politically empowered at national and international levels, it gradually had to respond to growing expectations regarding the focus of its activism. These expectations, originating both from within and outside the organization, demanded that the NBA increase its commitment to causes beyond the SSP and the Nimar valley. This option came at a cost.

Keeping the evolution of the Rondônia Forum and its member organizations as reference, I suggest that the same inverse correlation affected the level of activism and political empowerment of the NBA and its affiliated organizations, but the "arrows" ran in opposite directions. The gradual decrease of the role, visibility, and political power of the forum within and outside the Rondônia network, made way for—and was in part a result of—a growing capacity of its affiliated organizations to occupy that political space. In the case of the NBA, the organization's increasing role, visibility, and political empowerment both within the Narmada network and in Indian politics limited these possibilities for its member organizations. Local activism seems to have been locked within—and overly dependent upon—the catalyst potential of the NBA. In fact, local grassroots groups may have became the "losers" to the NBA's very success and of the subsequent pressures it endured to commit to goals beyond the Nimar valley.

The issue of technical capacity is key to understanding such limitations. It seems evident that the NBA eventually had difficulties coordinating simultaneous struggles at local, national, and international levels. To what extent was it able to systematically foment technical and political capacity among its member organizations while also focusing its resources on the larger battles that it fought after 1985? The (implicit) option of the NBA's leadership to focus the struggle on the more long-term and structural objective of challenging India's development model suggests that the organization eventually had to limit the scope of its local activism. This is not to say that the NBA and its key activists, such as Patkar, are no longer committed to seeing the struggle in the Nimar valley through to its final consequences. Many have already demonstrated their willingness to die supporting local populations. Their moral commitment to the local struggle is unquestionable. In political and strategic terms, however, the NBA's foremost commitment has become the struggle to influence India's development model. This option, however, may in part indicate an awareness of its own limits—and those of its local constituency—to resist the current backlash against the people affected by the SSP. Different from what one would have initially anticipated, the case of the Narmada network presents a strong challenge to standing assumptions in the literature on transnational advocacy networks about the role they play in empowering local populations.

This chapter concludes by qualifying one of its opening promises, that is, that the study of the Narmada network provides insight into the role of local groups in influencing a network's approach to environmentally sustainable development. In fact, the NBA's contribution to this issue is evident throughout the chapter. From the perspective of that organization, environmentally sustainable development is a process inherently related to as well as dependent upon social, economic, and political justice, and on concrete mechanisms for direct popular participation in policymaking arenas. Less evident, for the reasons discussed in the above paragraphs, is the contribution of grassroots groups in the Nimar valley to this approach.[57] It is certain that the plight of local groups has provided the baseline for the NBA's approach to environmentally sustainable development. There are few indications, however, that grassroots groups have devised concrete instruments for making this approach operational in the specific context of the Nimar valley. As such, the study of the Narmada network falls short in its value for a true understanding of what type of development local groups anticipate in the immediate future and what specific roles they expect to play in that process.

8

Conclusions

A transnational advocacy network's effectiveness in promoting environmentally sustainable development depends on the role that its local members have in framing network priorities and in devising and implementing its main strategies. The cases presented in this book reveal the internal politics of transnational networks' activism. They shed light on the dynamics and factors that foster—and hinder—local organizations' proactive role within transnational advocacy networks. In so doing, this study has challenged two existing assumptions, until now widely accepted by theorists and activists alike: that transnational actors, namely international and national non-government organizations (NGOs), are the key players in transnational environmental advocacy networks, and that participation in such networks inevitably empowers local groups. In fact, local groups are the ones who hold the key to a network's effectiveness and, unfortunately, their mere participation in transnational advocacy efforts does not necessarily lead to their empowerment. Such an outcome depends upon the process of "localizing" a network's activism and on how this process affects local politics, beyond the narrower confines of network politics per se.

For the last two decades, the phenomenon of transnational advocacy networks has been perceived, particularly in the areas of environment and human rights, as a positive development. Transnational networks have brought a variety of new actors, essentially representatives of the interests of different sectors of international and domestic civil societies, into international and domestic policymaking. Thus, they have contributed to an increase in democratic participation in these processes. In theoretical terms, transnational advocacy networks are helpful methodological tools in multilevel analyses. They contribute both to a better understanding of the role that actors of different natures play in policy design and implementation at local, national, and international levels, and to evaluations of the impact of such policies on populations and on socioeconomic and political structures at these different levels. This is of no small value in an increasingly interdependent world, one in which actors, processes, and policies at once affect and are affected by factors that cut across national borders and challenge traditional notions of space and time. Yet as the concept and the activism of transnational advocacy networks approach maturity in global politics critical evaluations of these instruments become all the more relevant.

This study has demonstrated the importance of evaluating the effectiveness of transnational advocacy networks, those addressing environmental issues in particular, not only in terms of their impact on national governments, international organizations, and the global public opinion, but also on the local human and natural environments. The focus on the local level revealed that transnational networks tend to be more successful in affecting policies and institutions at the international and national levels than at the local level. This is not only problematic for local groups participating in these networks, but sometimes even threatening to the material and physical security of individual activists.

The political and technical empowerment of local organizations is essential to guarantee that network goals are accomplished locally. Different from what has been widely assumed, however, the mere participation of local groups in transnational activism does not lead to their empowerment. This outcome may or may not occur depending on a network's internal politics, distribution of resources, choice of strategies, and selection of priorities—for instance, how its members formulate and pursue measures leading to environmentally sustainable development. The process of "localizing" a network's activism has direct consequences for local politics. It is here, perhaps, that this book provides its main contribution. While addressing the advantages that transnational advocacy networks provide to local partners in terms of resource sharing and mutually reinforcing strategies, it highlights the limitations of transnational networks—at least as they have been conceived until now—for the institutionalization of local activism and the long-term empowerment of local groups. In the following section, a brief comparison between the different phases of the Rondônia network, and between it and Ecuador's anti-oil and India's Narmada networks, stresses some critical aspects of this process of localizing a network's activism and the challenges that it encounters. At the end of this chapter, I present some policy suggestions for those directly involved in transnational environmental advocacy efforts and indicate directions for future research.

Local Empowerment and Local Results

Rondônia was not even a state when environmentally concerned individuals and organizations in Brazil and abroad mobilized to protect the region's environment. Local civil society organizations were practically nonexistent in Rondônia in the early 1980s. This situation reflected the transient nature of the state's civil society, composed mostly of recently arrived migrants, with no roots to the region or understanding about its sociopolitical and ecological dynamics. Traditional populations in the area, such as indigenous groups and rubber tappers, were essentially nonentities in the local political spectrum due to their low levels of organization. In this context, transnational mobilization against Polonoroeste's environmental impact unfolded—geographically and politically—outside Rondônia.

The absence of institutionalized local spokespeople (despite the contribution of individual activists residing in the state) aggravated certain divisions that plagued the Rondônia network in its formative years. The most significant of these referred to the network's main goals and arenas of activism. Without interlocutors who could assess the local merits and consequences of—and mediate between—the different priorities of national and international network members, activism lacked general coordination and resources were dispersed.

In the early years of the Rondônia network, its international members used the environmental devastation caused by the Polonoroeste project as exemplary of the need for an increased level of accountability by multilateral lending institutions for the environmental and social consequences of their development projects. International environmental NGOs and activists directed network resources toward pressure strategies in international arenas, such as the World Bank, the U.S. Congress, and the European parliaments. Brazilian activists, however, were first concerned with the impact of the project on Rondonian Amerindian populations and their environment, and second with the environmentally and socially unsustainable policies of the military regime, then in power, for Amazonia's development in general. While lending resources to international strategies, Brazilian activists resented the network's international focus. They attributed to it the loss of many opportunities for influencing implementing agencies in charge of Polonoroeste's environmental and Amerindian components. It is not surprising that the Rondônia network's most significant impact during the 1980s was on the World Bank. Activism against Polonoroeste became a cause célébre within the MDB campaign, and was instrumental in leading the bank to reformulate some of its policies and lending priorities. In Rondônia, however, deforestation, unsustainable agriculture, invasion of Amerindian lands and of conservation units remained unabated.

Mindful of the constraints that the absence of a local membership base imposed on network activism, the members of the Rondônia network invested significant resources in building such a base in the early 1990s. The imminent signing of the Planafloro project, a follow-up project to Polonoroeste, gave momentum to such efforts. In addition, Brazilian transition from authoritarian to democratic rule in the mid-1980s facilitated local mobilization in Rondônia. The creation of the Rondônia Forum in 1991 was the most significant consequence of this combination of a favorable political environment and the influx of external technical and material resources for local mobilization. The forum's mandate was to serve as a clearinghouse for initiatives by civil society organizations toward participation in Rondonian environmental and development policies. The forum's existence was, in itself, an asset to the Rondônia network. It constituted a valuable formal medium through which local grassroots groups could voice their needs and expectations vis-à-vis the network's initiatives. The forum was also welcomed by actors outside the network, such as the World Bank. The forum's involvement in Planafloro represented an opportunity for World Bank officials to stress their institution's commitment to participatory initiatives.

Through the early 1990s, despite the claims by the Rondônia Forum's leadership that its primary commitment was to the interests of its constituent members, namely, grassroots and advocacy NGOs in Rondônia, the forum remained primarily responsive to the agendas of the national and international members of the Rondônia network. The low levels of technical and political capacity of local groups as well as divisions between groups (for instance, between rural workers and Amerindians, who competed for the same key resource, land) limited their contribution to the network's activism. In hindsight, it was only natural that the forum's leading activists and organizations strengthened cooperation with network members that could provide them with the largest amount of support, namely, international environmental NGOs and their national allies in Brazil. These actors had been greatly empowered by previous successes of the MDB campaign and by favorable political circumstances in Brazil, such as higher levels of public environmental awareness as a result of the United Nations Conference on Environment and Development held in Rio de Janeiro in 1992.

Under the leadership of international environmental NGOs such as the Environmental Defense Fund, and strengthened in its legitimacy by the forum's existence, the Rondônia network achieved some of its goals, namely, the redesign of Planafloro in terms that were more environmentally sustainable than those of the project's initial version, and the selective incorporation of civil society organizations in the project's decision-making and implementing institutions.

As important as these conquests were in conceptual terms, they remained "on paper." They had no concrete effect on Rondônia's development path. As I have argued in previous chapters, the Rondônia network remained ineffective at the local level due to the weakness of its local membership base. The forum's formal existence, rather than attenuating this problem, aggravated it. As it struggled to participate in Planafloro and to keep the Rondonian government committed to the project's environmental goals, the forum relied increasingly on its international partners and emulated their activism. As a result, it unintentionally distanced itself from its affiliated organizations, with dire consequences for its own legitimacy within and outside the Rondônia network.

The forum's legitimacy crisis, culminating in 1994, represented a turning point in the evolution of the Rondônia network. It generated opportunities for revision of network strategies and priorities. It also forced the forum to reassess its institutional identity and commitments, a process that brought the organization's leadership closer to its members and to the populations they represented. An unprecedented level of political cohesion among local civil society organizations emerged from these processes, leading to a natural rise in their political assertiveness vis-à-vis other network members, the Rondonian government, and World Bank officials. In this atmosphere, the proposal by international environmental NGOs to take the Planafloro project to a newly created international grievances mechanism, the inspection panel, fell on fertile ground.

The success of such a strategy in terms of its concrete gains for Rondônia's environment, the reactions it generated from the Brazilian and Rondonian governments and the World Bank, and the political visibility that it granted to local groups in domestic and international arenas, further empowered Rondonian organizations. The restructuring of the Planafloro project and the formulation of the Program for the Support of Community Initiatives were among the most important consequences of these parallel processes of local network members' internal reassessment of priorities and consensual decision to embrace an innovative strategy.

One of the most interesting aspects of the story of the Rondônia network starts where many assumed it was close to the end. The Planafloro restructuring process and the "upper hand" that local civil society organizations had in its outcome evidenced the level of political empowerment that these groups had achieved as a result of their participation in a transnational advocacy network. But the project restructuring was not the end of the story. Local groups' political empowerment shifted the balance of forces within Rondonian politics and among the actors that participate in local environmental and development policymaking.

It is here that transnational activism acquires implications still not fully addressed by theorists and activists alike. In the three cases evaluated in this book the political empowerment of local groups as a result of their participation in a transnational advocacy network had immediate consequences. Most network members, however, were unable to envision the possibility of such consequences having negative impacts (as well as positive ones). As such, they were unprepared to minimize such negative impacts. In the case of Rondônia, the very success of the network's activism brought upon local groups new responsibilities for the formulation, implementation, and monitoring of policies that they were not always prepared to shoulder. In addition, the concrete environmental gains achieved by the network in the context of Planafloro generated a political backlash on the part of sectors of the government and of local economic and political elites. In recent years, opposition forces have reiterated efforts to roll back some of the environmental gains obtained by the Rondônia network.

In India, the backlash against local groups has been even more pronounced than in Rondônia. The Narmada network has obtained an unprecedented level of success both at the international level, influencing changes in international organizations, and within India, where it has formulated solid critiques of the nation's democracy and development model. Yet network activism has been unable to stop the flooding of villages in the Nimar valley and to shelter villagers and displaced populations from state repression. In part, these difficulties can be attributed to the incapacity of the Narmada Bachao Andolan, the main catalyst for local activism, to cope with the demands and expectation placed upon it (at local, national, and international levels) as a result of the Narmada network's successes throughout the 1980s and 1990s.

In Ecuador the challenge to the process of "localizing" transnational activism has specific nuances. Several arenas for local groups' participation in decisions about oil development in the *Oriente* have emerged as a result of the network's activism, combined with the growing assertiveness of the country's indigenous movement throughout the 1990s. Yet the level of capacity for a meaningful participation in these arenas varies greatly among local groups. An unexpected challenge to local activism is the downsizing of oil operations by large corporations and their replacement by smaller, less well-known companies spread throughout the *Oriente*. While transnational activism against oil giants such as Texaco and ARCO is not the only factor explaining such a trend, it definitely has had an impact in the declining level of interest by "brand name" corporations in conducting oil activities in Ecuador's Amazonia. But shouldn't the diminishing interest of corporate giants in oil exploitation in Ecuador be considered a major achievement of the anti-oil network? Yes and no. Yes if one assumes that oil development in the *Oriente* may slow down and be restricted to smaller areas due to the more limited financial and technical resources of smaller companies when compared to those controlled by the oil giants. No, if one realizes the difficulties that local activist groups may face in gathering information and monitoring the activities of several smaller companies, which are often unconcerned with the potential impact of environmental activism on their corporate image.

The fact that activism in transnational advocacy networks often generates unexpected political and technical challenges to local civil society groups should not overshadow the merits of such initiatives. For instance, despite political backlashes and technical hurdles, participation in the Rondônia network has created conditions for local groups to formulate their own approach to environmentally sustainable development. Such an approach was fully developed in the context of the Program for the Support of Community Initiatives. This approach has already produced a positive impact on Rondonian human and natural environments as selected PAIC projects have fulfilled their objectives. In light of these findings, I risk a prediction: in the (near) future, the approach of Amazon groups to environmentally sustainable development—a process that integrates environmental protection with the improvement of communities' socioeconomic well-being and political participation—is likely to prevail in policy initiatives in Rondônia.

The role of local groups in formulating an integral approach to the concept of environmentally sustainable development, one that considers environmental protection a process inherently linked to the improvement of populations' material and political conditions, is a common trend among all the networks studied in this book. If one accepts the argument that the effectiveness of a transnational advocacy network at the local level is a function of the role that its local membership base plays in the mobilization, then one must also accept that local environmentally sustainable development will only occur when defined and implemented according to local groups' visions and priorities.

Indigenous and settlers' populations in Ecuador's *Oriente* have attempted to formulate a vision of what environmentally sustainable development means for a region endowed with significant natural resources, crude oil among them. They have succeeded to different degrees. Political participation of grassroots groups in larger development policies for the *Oriente* is a notion that is now accepted, at a rhetorical level at least, by most sectors of Ecuadorian politics. It is still not clear, however, how this issue will evolve and to what extent grassroots groups may be able to strengthen their role in regional policymaking. What has become evident is that the activism of Ecuadorian groups at the national and local levels within the framework provided by the anti-oil network has generated spaces for dialogue and consultations among government officials, oil companies, and affected populations. This dialogue has created opportunities for the formulation of instruments for the compensation of affected populations and for further research on the area's ecological characteristics, among others.

Despite the gains that local populations may obtain through the formulation of a consensual approach to environmentally sustainable development, one should remain mindful of the complexity inherent to this process. Such a complexity has remained a threat to the effectiveness of the anti-oil network's efforts. It risks undermining the unity among claimants in the lawsuit against Texaco and between claimants and their supporters (national and international environmental groups). In the case of the campaign against ARCO, the challenge is even more striking. Despite years of negotiations with the company, indigenous organizations representing the affected population have been unable to formulate a common plan for the company's support of a socioeconomic development plan for the region.

In India, the contribution of the Narmada Bachao Andolan to the formulation of a powerful critique of the country's development model is undeniable. Most important yet is the contribution that the NBA and its allies have made to the process of reevaluating the role of large hydroelectric projects in global energy policies. One concrete result of such a contribution was the commissioning of the work of the World Commission on Dams and the publication of its report in 2000. Another is the ongoing discussion in India about a national policy for resettlement and rehabilitation of populations affected by large dams. Unfortunately, this research was unable to identify (or isolate) the specific role that grassroots groups in the Nimar valley have had in these processes. This difficulty is in part related to both the nature of local groups' organization and of their insertion in the Narmada transnational advocacy network.

The comparison between the Rondônia, the Ecuador anti-oil, and India's Narmada networks reveals the importance of formal channels for the organization of local groups and for their insertion in a network's activism. In Rondônia, this role was performed by the Rondônia Forum. In Ecuador, regional and national indigenous federations and confederations and national environmental groups were instrumental. In the case of the Narmada valley, the Narmada Bachao Andolan is undoubtedly the main catalyst for local mobilization. While the existence of umbrella organizations or clearinghouses are essential to structuring local groups' participation in transnational advocacy networks, the nature of such organizations varies widely and so does the type of mediation that they perform between the interests of their affiliated groups and those of other network members.

The Rondônia Forum was essentially conceived and made operational by international and Brazilian NGOs involved in the MDB campaign. While it was welcomed by local groups, they remained in the background of initiatives during the forum's formative years. Eventually, the forum's leadership—individuals and local NGOs closer to domestic and international network members—assumed a proactive role in the Rondônia network. Such a role, however, did not always translate the level of commitment or engagement of other forum-affiliated organizations to the network's priorities and strategies. Inevitably, this situation led the forum to an "identity crisis" that affected its legitimacy as a valid interlocutor of local groups' interests within and outside the Rondônia network. As Rondonian civil society groups resolved the forum's legitimacy issues and reassessed its mission, they changed the nature of their engagement in the Rondônia network. In becoming proactive players within the forum, grassroots groups such as the Organization of Rondonian Rubber-Tappers and the Federation of Agricultural Workers of Rondônia came to have a decisive voice within the network as a whole. This process reflects a dual dynamic: participation in the forum led to a gradual increase in grassroots organizations' capacity; and their increased capacity eventually constrained the forum's role. At the turn of the millennium the forum has become less a catalyst organization for local groups' activism and more a source of resources to support grassroots' initiatives. This change is significant in that it allows a clear assessment of the role of local groups in transnational advocacy networks.

The role played by grassroots groups in shaping strategies and defining the priorities of the Narmada network, on the contrary, is less evident. The extraordinary success of the Narmada Bachao Andolan as a catalyst for local groups' activism may have hindered these groups' ability to develop a more proactive behavior within the network. As a consequence, it is difficult to distinguish between the contribution of the NBA's leadership and that of its individual affiliated organizations. This should not lead one to infer that the NBA is any less committed and accountable to the interests of its affiliated groups than the Rondônia Forum, for instance. The issue here is how the organization has mediated between the priorities of its very diverse constituency and how this process has impacted on the evolution of the Narmada network. Data suggest that the NBA has been more responsive to concerns of groups that prioritize changes in environmental and development policies that are national in scope than to the demands of groups that have a narrower, localized agenda. This situation is at least partially explained by the limited capacity of local groups (who do not seem to have been empowered by their participation in the Narmada network in the same proportion that the NBA has) and by the Andolan's need to focus its efforts on a subset of issues to avoid overstretching its resources.

The nature of the NBA's mediation between local grassroots groups and other members of the Narmada network is further illuminated by contrasting it to the mediation performed by Ecuadorian national organizations in the context of the anti-oil network. While the relation between the Andolan and the Narmada network's international partners was clearly defined, with the NBA asserting its financial and political autonomy and leadership role in determining the network's priorities and strategies, the nature of the organization's relations with its national and local partners is less evident. This is in part due to the Andolan's own ambivalence vis-à-vis its identity, that is, whether the organization is a grassroots one, representing the interests of specific populations, or whether it is a coalition or umbrella organization that mediates between the interests of different groups and facilitates their activism.

In Ecuador, on the contrary, the relationship between local grassroots groups, national organizations (indigenous federations and environmental NGOs), and their international partners has unfolded according to clearly defined patterns of interaction at all levels. Local indigenous groups, for instance, formally declare their affiliation to regional and national indigenous confederations and give them a mandate to represent their interests at the national and international levels. In local matters, however, these groups tend to assert their autonomy. The nature of relations between groups at different levels is even better defined in the case of local indigenous and settlers' groups, and their links with national environmental NGOs. Partnership is defined in very specific terms and environmental groups were never given a mandate to represent local populations. National NGOs and indigenous confederations do assume the largest part of the responsibility for mediation between local and international interests, with different degrees of success. National organizations are thus key elements in the process of resource sharing that characterizes transnational networks. For instance, in the anti-oil network, they were the main channels of articulation between local and international actors within the Amazon Coalition. The coalition was created with the goal of institutionalizing processes of sharing and exchanging political, technical, and material resources among actors concerned with environmental and indigenous issues in Amazonia. It has fulfilled its mission well and avoided challenges to the legitimacy of its initiatives to a large extent because its members have been mindful of the boundaries that constrain the actions of actors operating at different levels.

The members of the Narmada network, more specifically the Narmada Bachao Andolan, however, are not the only ones plagued by the porous boundaries of multilevel activism. Compared to the success of Ecuador's anti-oil network in this area, the shortcomings of the Rondônia network are even more striking. In the twenty years in which the Rondônia network remained mobilized, there were very few occasions when national actors actively participated in its efforts. For the most part, the Rondônia Forum played the role of mediator between the interests of international and grassroots groups, not always successfully. Activism in Rondônia was characterized, from the early 1990s on, by direct interactions between international and local organizations, in complete disregard for the potential advantages of mediation by national advocacy organizations. As a result, the process of resource sharing among network members was constrained. This has had negative consequence for local groups, particularly in the years following the restructuring of the Planafloro project. As the network's priorities became essentially local, international interest in it decreased. The resources once available to Rondonian groups from their international partners have since diminished. Historically, the lack of a systematic participation of national organizations has prevented the Rondônia network from institutionalizing avenues for resource sharing between local and national activist groups. This partially explains the difficulties that local groups have had in addressing issues related to the low technical capacity of their cadres (national organizations could play an important role in efforts to increase the technical capacity of Rondonian organizations—as they have at specific junctures, such as during the negotiations for Planafloro's restructuring, for instance). Yet the failure of local groups to reach out to and participate in national arenas for civil society environmental and development activism has perpetuated this problem. The irony is that part of the explanation for why local groups have failed to strengthen links with their national partners in Brazil is their limited capacity. They have lacked human and financial resources (such as competent cadres who understand issues debated in national meetings and financial resources to send representatives to conferences and rallies) that could enable them to fully engage in national activism. The consequence is the perpetuation of a vicious cycle in which lack of engagement and closer cooperation with national groups prevents gains in local groups' technical capacity, and limited capacity constrains local-national cooperation.

The paragraphs above summarize the commonalities and relevant differences among the transnational advocacy networks discussed in this book. In the next section I highlight certain lessons and recommendations that have emerged from this comparative study. My hope is that they may be valid contributions for future efforts to promote local environmentally sustainable development and effective empowerment of local civil society groups throughout the world.

Lessons and Recommendations

I. The formation or identification of local umbrella organizations or clearinghouses to support local groups' organization and mediate between them and national and international network partners should be a priority for activists. In light of the experiences narrated in this book, all efforts should be made to guarantee that umbrella organizations remain a participatory forum where different local groups may reconcile their differences and reach consensus vis-à-vis their goals and expectations as network members. In addition, the leaders or members of an umbrella organization's executive secretariat should remain mindful of the importance of not overshadowing the autonomy of local groups. Umbrella organizations should be arenas for resource sharing between network members and for the facilitation of members' initiatives. They should not control such resources or be the sole catalyst or initiator of local activism.

II. The existence of local umbrella organizations should be no substitute for well-established lines of cooperation between network members at all levels. Members of a well-structured transnational advocacy network, while engaging in activism at local, national, and international levels, should be mindful of the political boundaries of their actions. While such concerns may increase the burden of communication and information diffusion among members and make network decision-making processes slower and more cumbersome, they solidify a network's structure and integration. In the long term, this may benefit efforts to increase local groups' technical and political capacity beyond the specific confines of network activism (Ecuador's anti-oil network is exemplary of this possibility).

III. Both the establishment of local umbrella organizations and the processes of resource sharing and information diffusion typical of transnational advocacy activism are likely to affect the balance of power among local political actors. Network members at all levels should be mindful of the consequences of this process, particular for local civil society organizations and activists. This issue should be of particular concern for international network members. The cases discussed in this book demonstrate that transnational activism inevitably affects local politics. To what extent are international network members willing or able to become involved in domestic and local politics? What are the consequences for such actors' legitimacy and accountability to their global constituencies? To what extent do their resources allow them to commit to struggles that are inherently long-term, since they involve structural change? How may international network members best cope with the political responsibility of committing to transnational socioenvironmental activism and campaigns?[1] These issues should be fully addressed by network members, preferable at the onset of a given mobilization. The risk here is the tendency of international NGOs and activist groups to prioritize action on immediate crises, which prevents their long-term involvement with processes of strengthening local civil society organizations. This behavior may be explained by several factors: 1) the Anglo-Saxon "work ethic" that pressures organizations to "move on with business," that is, to define goals and accomplish them in the shortest possible time and at the lowest cost; 2) Northern groups' pursuit of accountability and responsiveness to their constituencies and their expectations of a "successful" ending to a given mobilization effort or "crisis response"; 3) bureaucratic and time constraints imposed on Northern activist groups by large donors and funding sources.

IV. Another challenge to the relationship between international and national/local network members refers to their choice of strategies. Given the superior resources of international groups, transnational networks have had a tendency to concentrate advocacy efforts on international arenas. Yet this practice generates at least two risks for a network's effectiveness. First, successful strategies, conceived and implemented by international NGOs in industrialized countries, do not necessarily fare well when reproduced by activist groups in the South. The latter should evaluate the potential impact of emulating Northern NGOs' strategies and the extent to which these may affect their own capacity. They should also be sensitive to the impact of certain strategies on their local political and cultural context. All networks studied in this book presented examples of the complexities and potential for internal divisions entailed by the choice of inserting local activism into preestablished international leverage mechanisms such as institutionalized lobbying structures or campaigns. A second risk of an excessive reliance on international strategies is the possibility of its constraining the emergence of endogenous and often innovative channels for activism. A key example is the potential that local and national lawsuits have had for accomplishing specific network objectives. Yet, court actions have remained, at least for the three networks studied here, an occasional rather than a systematic channel for activism. Given the potential for long-term and structural change that this avenue entails (as indicated by concrete examples in Rondônia, Ecuador's *Oriente,* and India), network members should devote further material and intellectual resources to the consolidation or institutionalization of this route.

V. The priority to locally devised strategies within the context of transnational activism, and an emphasis on the institutionalization of legal measures as an avenue for the promotion of structural change are inherently linked to another key process, that of defining environmentally sustainable development. Network members must encourage the formulation of an approach to environmentally sustainable development that truly represents the needs and expectations of local network members. A network's strategies (legal, educational, political) and goals should remain faithful to such an approach. The essence of this process is to define what specific actions, projects, and policies must be pursued—at different levels—in order to foster environmentally sustainable development in a given region. Such actions' and policies' contribution to national and global environmental sustainability, while a desirable outcome, should remain secondary goals in the agendas of network members.

VI. Members of transnational environmental advocacy networks should be mindful of their potential role in redefining (or better defining) the concept of environmentally sustainable development. As I discussed in the Introduction, dominant approaches to the notion have underplayed the need for structural changes in processes aiming at environmental sustainability. The activism of local communities in Brazil, Ecuador, and India has demonstrated that if issues of distributive justice and meaningful political participation are not addressed, effective environmentally sustainable development will remain an elusive goal.

VII. Network members must be prepared to respond to demands generated by the success of their transnational activism. Part of this issue was addressed in item III above, where I highlighted the political implications of engaging in transnational advocacy networks. In addition, increased technical and material demands on network members are likely to result from a network's increased political assertiveness. Network members may be called upon to participate in the design of socioenvironmental policies, as was the case of Rondonian civil society groups and the formulation of the Program for the Support of Community Initiatives. They may also be requested to cooperate in environmental impact assessments, as OPIP did when it joined ARCO's technical environmental committee. It is important for all network members to devise policies for their long-term engagement in decision-making arenas, which may become available to them as a result of successful activism. Network members should be cautious against the temptation to embrace responsibilities beyond their capacities, or to accept tasks that traditionally fall under the competence of the state or oil companies, for instance. While NGOs (domestic and international), research institutes, and grassroots groups should be open to cooperate with state agencies and corporations, they should limit their role to independent advising. In very well-defined and limited cases, network members may assume implementation responsibilities for small, locally based initiatives. By assuming large-scale executive roles in environmental and development projects and policies network members may be misidentified by their constituencies as being in charge of delivering goods and services. When such goods and services are not provided, network members may see their legitimacy challenged.

Final Thoughts

Shedding light on the scope of action of transnational advocacy networks and on their internal processes of mediation among actors operating at local, national, and international levels is at least one way of addressing the limitations of existing theories on international environmental relations and global interdependence. As it has been noted elsewhere,[2] existing theories lack the capacity to fully address the role that actors of different natures play in processes and policies that unfold and have simultaneous consequences at local, national, and global levels of analysis. Problems such as environmental degradation have implications at all of these levels, affect stakeholders at all of these levels, and demand solutions at all of these levels. Yet neither theories that emphasize the importance of the nation-state for global environmental management nor those that emphasize international institutions—such as regimes, treaties, or international organizations—have the breadth to address all the variables that affect such a process. This is simply because most approaches are unable to account for the impact of local politics on national and global processes. The study of transnational advocacy networks from the inside out is a valid contribution to efforts toward integrating domestic and international analytical realms.[3] Its originality, however, lies in "deepening" the analysis of domestic variables, highlighting the relevance of local actors and local processes for global politics on the one hand, and, on the other, forcing analysts and policymakers alike to consider the implications of domestic and international initiatives for local populations, local civil societies, and the local environment.

Despite such contributions, the study of transnational advocacy networks is still wide open to further inquiry. Evidently, this book has not answered all the questions that the topic raises, either in empirical or in theoretical terms. While it has drawn attention to the role that local actors play in transnational efforts to promote local and regional environmentally sustainable development there is still room for further research on the contribution of local approaches to the formulation of national and global environmental protection initiatives. We do know that abstract and general definitions of environmentally sustainable development, while often generating consensus among national and transnational actors, have done little to promote that kind of development in practice and particularly at the local level. To what extent may inductive approaches to the concept fare better?

Finally, many of the arguments made in this book should benefit from future studies that assess their validity beyond the realm of socioenvironmental policies. Transnational activism in the areas of human rights, women's rights, and the rights of refugees and populations in exile should provide interesting material to enhance one's understanding of the role of non-state actors, and particularly of those operating at the local level, in dynamics and processes that have national and international implications.

NOTES

CHAPTER 1. INTRODUCTION

1. Interview, 23 May 2000, Rondônia.

2. For a discussion on theoretical trends in global environmental governance see Paterson (1999).

3. The term *transnational environmental advocacy network* derives directly from the work of Margaret Keck and Kathryn Sikkink (1998). The authors' object of study are "transnational advocacy networks," which are evaluated in the context of environmental, human rights, and women's issues. These issues, however, do not exhaust the universe of transnational advocacy networks, which in theory can form around many other issues such as trade and health. This book presents a study of transnational *environmental* advocacy networks. While many of its conclusions apply, specifically, to these types of networks, its most important lessons contribute to improving understanding of transnational advocacy networks in general.

4. Keck and Sikkink 1998, 2.

5. Ibid., 4.

6. Keck and Sikkink 1998; Princen and Finger 1994.

7. Rich 1994; Fox and Brown 1998.

8. Lipschutz 1996; Wapner 1996; Brysk 2000.

9. Princen 1994; Jordan and Van Tuijl 2000; Jezic 2001.

10. Princen and Finger 1994; Keck and Sikkink 1998.

11. Fox and Brown 1998.

12. Environmental protection measures may be local, national, and international in scope. Yet as Lipschutz and Conca (1993) state, it is of the very transnational nature of environmental issues to have implications on all three levels. Thus, for instance, international and macroeconomic measures to protect tropical forests eventually contribute to the preservation of specific forest reserves and indigenous lands, while local efforts to preserve the latter have global implications for the concentration of greenhouse gases in the atmosphere and availability of biodiversity stocks.

13. The term *local membership base* defines a subset of network members that are physically located at the site or region of primary concern to the network as a whole. The term does not necessarily imply the existence of large local constituencies or that these are politically strong.

14. Rosenau (1993) labels this phenomenon as "subgroupism." In a "multicentric" world (as opposed to a "state-centric" world), subgroups, NGOs, and multilateral organizations gain importance vis-à-vis nation-states in determining global processes and structures.

15. For the purposes of this study, "actor" or "political actor" is an entity that participates in political life (Frey 1985) and engages in power relations with other actors (Young 1972). The term is usually applied to a group of individuals who share common characteristics, interests, and behavioral cohesion (Fox 1992).

16. Boissevain 1974.

17. Heclo 1978.

18. Haas 1989; Sikkink 1993; Keck and Sikkink 1998.

19. Jordan and Tuijl (2000) also question, implicitly, the notion that values and principles determine behavior in a network. These factors are excluded from their redefinition of transnational advocacy networks, which is stated as "a set of relationships between NGOs and other organizations that simultaneously pursue activities in different political arenas to challenge the status quo" (p. 2053).

20. Other network members are, usually, individual activists, church-based organizations, media channels (newspapers, specialized journals), and sectors of governmental or multilateral agencies (sometimes staff in a specific division or department).

21. Vakil 1997, 2060.

22. McCormick 1993, 132.

23. Rodrigues 2000, 128.

24. Keck and Sikkink 1998, 6–8.

25. O'Donnell and Schmitter 1991; Migdal et al. 1994; Jelin and Hershberg 1996.

26. Schmidt 1997, 13.

27. The notion that national civil societies have grown stronger in recent decades is controversial. There is a general consensus on the issue when one looks at countries that have moved away from authoritarian and totalitarian regimes and toward democratic governments. Yet some question the extent to which the expansion of market capitalism and the reduced role of the state have made individuals strong as consumers, but weak as citizens ("A Poor Case for Globalization," by Philip Stephens, in *Financial Times*, August 17, 2001).

28. This is not a consensual notion, particularly in the realm of international relations. O'Brien et al. (2000) summarize the controversy, explaining that traditional international relations literature speaks of a "society of states," and doubts the existence of a global civil society in the absence of a global state (p. 13).

29. Castells 1996.

30. Keck 1995.

31. Brysk 2000.

32. O'Brien et al. 2000, 13.

33. INZET Association 1999.

34. On the issues of civil society as an arena of conflict and the interpenetration of state and civil society, see Wapner 1996 and O'Brien et al. 2000.

35. Wapner 1996, 4. Alternatively, Lipschutz (1996 and 1997) equates the concept of civil society to that of a "regime," composed of local, national, and global NGOs. This is a more restricted definition of global civil society than most, but still useful to provide a context for transnational advocacy networks.

36. Mohan and Stokke 2000; Pretty and Ward 2001.

37. This has been a recurrent criticism of Keck and Sikkink's approach to transnational environmental networks. See for instance Jordan and Tuijl 2000 and Rodrigues 2000.

38. Lipschutz 1997, 94.

39. The "fuzziness" of the concept of environmentally sustainable development has been a common complaint in the literature; see for instance Lélé 1991 and Kothari 1995.

40. The WCDE Report defines (environmentally) sustainable development as "development which meets the needs and aspirations of the present without compromising the ability of future generations to meet their own needs" (p. 43).

41. Many have criticized the term as a "contradiction in terms" (O'Riordan 1985) or a "development truism" (Redclift 1987). But it is Lélé (1991) who best spells out the "lack of consistency" in interpretations of sustainable development, critically reviewing the literature and suggesting alternatives for the clarification and better articulation of the concept.

42. I borrowed the term from IBASE (1995). Lélé (1991) offers an alternative for the label: "ecologically sound and socially equitable development."

43. Archibugi, Nijkamp, and Soeteman 1989, 3.

44. Norgaard 1984.

45. Pearce and Warford 1993.

46. Natural Capital Stock is "the stock of all environmental and natural resource assets, from oil in the ground to the quality of soil and ground water, from the stock of fish in the oceans to the capacity of the globe to recycle and absorb carbon" (Pearce et al. 1990, 1).

47. Economists at the World Bank have worked to devise new ways of including environmental value in traditional economic calculations (Interview with Gunar Eskelan, World Bank economist, Washington, D.C., March, 1991). One outcome of this

effort has been a new ranking of countries that underplays the traditional criterion of GDP per capita and emphasizes countries' environmental resources as sources of wealth.

48. Lélé 1991.

49. Dalton 1993.

50. Lélé 1991; Milbrath 1993.

51. Lélé (1991) explains that mainstream sustainable development approaches seem to "have quietly dropped these terms [. . .] and instead focused on 'local participation'" (p. 614).

52. Adams 1990, 83.

53. Blaikie 1985; Adams 1990.

54. Adams 1990; Shiva 1991a and 1991b.

55. Shiva 1991a.

56. In Hall (2000), for instance, "'sustainable development' [in the Amazon context] is a matter of conserving both biological diversity (biodiversity) and social diversity (sociodiversity) as two complementary sides of the same coin" (p. 2).

Chapter 2. The Dilemma of Amazonian Development and Its Impact on Rondônia

1. Mahar (1989) explains that there are two geographical concepts of Amazonia used in Brazil: Legal Amazonia, comprising seven states in the northern region of Brazil and parts of two central-west states, is used by the government for regional planning and policies; and Classic Amazonia, comprising only six states and coinciding with Brazil's northern region, is used for statistical purposes. The concept adopted in this research is that of Legal Amazonia.

2. Moran 1981.

3. Mahar 1989. For comparative purposes: Northern forests contain about ten to fifteen species of plants; the number of fish species in the Amazon Basin is eight times the number found in the Mississippi River system.

4. It is estimated that the mineral deposits in Carajás contain 18 billion tons of iron ore, 25 million tons of nickel, 60 million tons of manganese, 1.2 billion tons of copper, as well as significant quantities of cassiterite, bauxite, and gold (Fonseca 1981).

5. The Brazilian Superior School of War (Escola Superior de Guerra—ESG).

6. Ernesto Che Guevara, Fidel Castro's main political and military advisor, led guerrilla forces in Bolivia and was eventually killed by the army in that country on October 9, 1967.

7. Respectively, the *Superintendência para o Desenvolvimento da Amazônia* (Superintendency for the Development of Amazonia—SUDAM), and the *Banco da Amazônia* (Bank of Amazonia—BASA).

8. Mahar 1989.

9. IBASE 1990, 7.

10. Hecht 1985, 370.

11. Binswanger 1990.

12. For comparative purposes, corporate agricultural profits were taxed at 6 percent while profits from other sources were subjected to tax rates between 35 and 45 percent (Binswanger, 3).

13. Bunker 1985; Hall 1989; Mahar 1989; and Binswanger 1990.

14. Hall 1989.

15. Mahar 1989.

16. Binswanger 1990.

17. IBASE 1990, 8.

18. Agro-industry Census, IBGE, 1970 and 1980 (in IBASE 1990, 10). The five Amazonian states are Acre, Amazonas, Roraima, Pará, and Amapá.

19. Mahar 1989, 6.

20. These projects were implemented through the following World Bank Loans: Ln 2062 (road), Ln 2061 (health services), Ln 2060 (colonization and environment), and Ln 2353 (new settlements). The Amerindian project was not directly funded by the World Bank.

21. The World Bank 1981.

22. Ibid.

23. Aufderheide and Rich 1985.

24. See Davis 1977; Treece 1987; and Price 1989.

25. *Folha de São Paulo* (newspaper), 23/3/89.

26. See for instance the series *"A Conspiração contra o Brasil"* (A Conspiracy against Brazil), published by the newspaper *O Estado de São Paulo,* from August 9 to 15, 1987. The "evidence" of a conspiracy was later proved to be false, but the series raised alarm in the Brazilian congress and affected the work of environmental and human rights groups in the country.

27. Kolk 1998.

28. Once again the thin administrative structure in the Amazon region prevented the implementation of environmental protection measures, while excessive bureaucracy paralyzed the fund's resource distribution processes.

29. The G-7 was composed of the seven most industrialized nations in the world, the United States, United Kingdom, Japan, France, Germany, Canada, and Italy. For details on the formation of an international consensus on the initiative see Kolk 1998.

30. Ros Filho 1994.

31. Demarcations have occurred even despite the political difficulties that the process entails in Brazil. See Moore and Lemos (1999) for an example of the political tactics used by conservative interests against the demarcation of Amerindian reserves.

32. Interviews with Brent Millikan, consultant for the Brazilian Ministry of the Environment, and ex-consultant for the Rondonian Forum of NGOs and Social Movements, and Roberto Smeraldi, Coordinator of Friends of the Earth—Amazonia Program. Respectively, Brasília, May 19, 2000, and São Paulo, June 5, 2000.

33. Federal and state counterpart funds for the PNMA never materialized and Brazil ended up paying high penalty taxes for not using the resources disbursed by the World Bank. IBAMA's own institutional fragility also contributed to the program's failure.

34. The state of Rondônia was not alone in this strategy. The state of Mato Grosso, for instance, also presented to the World Bank in the early 1990s a proposal for funding of the Mato Grosso Natural Resources Management Project (Prodeagro).

35. As I will discuss in the following chapter, in 1987 the World Bank underwent an organizational reform that increased its resources for environmental initiatives. The reorganization was in part a response to pressures from international NGOs to make the bank accountable to the environmental and social consequences of its development projects.

36. The objective of the Rondônia Zoning Plan is to identify soil characteristics and natural resources potential in different areas in the state of Rondônia to rationalize the state's economic development.

37. The World Bank 1992a.

38. Ibid., 31.

39. Oliveira Filho 1990.

40. Dreifuss 2000.

41. For details on the Yanomami campaign see Schmink 1992.

42. Hall 1989; Oliveira Filho 1990.

43. Hall 2000.

44. Plan Colombia is an initiative designed and financed by the U.S. government to help the Colombian government fight drug traffic and guerrillas. It has raised concerns related to sovereignty and regional political stability among many Latin American countries. Plan Colombia started in 2000.

45. In the year 2000, budget allocations for military programs in the area absorbed 44 percent of the total resources for the region while investments in infrastructure and development received 54 percent. Resources for environmental protection initiatives in Amazonia amounted to a mere two percent (*Observatore* [newsletter], Year III, number 31, May 2001, INESC, Brasilia, 7).

46. For a critical evaluation of *"Avança Brasil"* see Carvalho, Georgia et al. 2001.

Chapter 3. Urgent Action! Transnational Mobilization against Disaster in Rondônia

1. The World Bank 1992c, 29.

2. Price 1980; Maybury-Lewis 1981.

3. Price 1989, 164.

4. Two individuals, Barbara Bramble at NWF and Stephen Schwartzman at EDF, had years of professional experience in Amazonia and innumerable contacts with the Brazilian environmental and human rights community. But it was Bruce Rich, also from EDF and Brent Blackwelder, from the Environmental Policy Institute, that brought to the network crucial lobbying skills.

5. As Rich (1994) explains, the American development community, including environmental NGOs, had been, for too long, constrained in its capacity to produce and make available critical evaluations on certain topics. This was due, in part, to the fact that "many academics and research groups specializing in international development either had contractual relations with the World Bank, were partly dependent on it and other international agencies for income, or looked to these agencies for future employment" (p.113).

6. The interest of Northern groups in development and environmental problems in developing countries has often raised issues of sovereignty and imperialism. One way for Northern NGOs to circumvent such concerns is to establish close partnership with Southern groups who endorse their activism.

7. As the campaign unfolded, European and Southern environmental and human rights organizations also joined in.

8. Interview with Bruce Rich, lawyer for EDF, Washington, D.C., March 13, 1991.

9. "Statement of Bruce Rich on behalf of Environmental Defense Fund and National Wildlife Federation concerning the Environmental Performance of the Multilateral Development Banks and the International Monetary Fund, before the Subcommittee on Foreign Operations Committee on Appropriations United States House of Representatives, April 24, 1989" (p. 1).

10. Rich 1985, 1987, 1990, 1994; Schwartzman 1988.

11. *The New York Times,* October 10, 1986.

12. Hearing before the House Science and Technology Subcommittee on Natural Resources, Agricultural Research and Environment, September 1984.

13. Leading newspapers such as *The New York Times, The Washington Post, Financial Times, Milwaukee Journal,* magazines and other periodicals, such as *The Economist, News & Comment, US News & World Report,* as well as environmental journals, *The Ecologist,* and *Conservation Biology,* published articles on the MDB Campaign or on the issues it focused on.

14. The World Bank 1994.

15. Rich 1994; Interviews with Rich and with David Maybury-Lewis, anthropologist, professor at Harvard University, and World Bank consultant for the Polonoroeste, Cambridge, March 3, 1991.

16. Environmentalists' testimonies in the 1985 and 1986 Hearings on Foreign Assistance and Related Programs Appropriations.

17. Interview with Rich and Rich 1994.

18. EDF's Polonoroeste Information Package (mimeo).

19. Interview with Rich, and EDF's (n/d): "Chronology of Events" (mimeo).

20. Rich 1994.

21. Interviews with Ana Maria Avelar, President of the Rondonian NGO *Instituto em Defesa da Identidade Amazônica*, INDIA, Rondônia, November 24, 1994, Betty Mindlin, Brazilian consultant for Polonoroeste, São Paulo, September 12, 1994, and Eduardo Martins, environmental consultant for Polonoroeste, Brasília, September 28, 1994.

22. The southern part of the region encompassed by the Polonoroeste project is characterized by a savanna type of vegetation, known in Brazil as *"cerrado."*

23. Interview with Martins.

24. Interviews with Robert Goodland, World Bank Senior Environmentalist, Washington, D.C., July 19, 1993, Mindlin, and Azis Ab'Saber, Brazilian geographer and consultant for the Carajás Iron Ore Project, São Paulo, September 8, 1994. These individuals stressed that current environmental knowledge on the region has been achieved at a high cost to Amazon physical and human environments.

25. See, for instance, Binswanger 1990; Hecht and Cockburn 1989; and Mahar 1989.

26. McCormick 1989.

27. See note 12 for examples of media participation in disclosing information on Polonoroeste's environmental effects.

28. Interview with Mindlin.

29. Pádua 1990.

30. Namely, Robert Goodland.

31. To meet the expansionary demands of its agricultural division, the Bank pushed for Polonoroeste to become an integrated settlement program based on agricultural and agroforestry activities.

32. The World Bank 1992c; Freitas and Soares 1994.

33. Davis 1977; Maybury-Lewis 1981; Price 1989.

34. Price 1980; Maybury-Lewis 1981; interview with Goodland.

35. Interviews with Mindlin; José Juliano de Carvalho Filho, Coordinator of FIPE's Polonoroeste monitoring team, São Paulo, September 9, 1994; and Marita

Kock-Weiser, World Bank staff and member of the Polonoroeste management team, Washington D.C., July 21, 1993.

36. The World Bank 1992c, 93; interviews with Luis Coirollo, World Bank Task Manager for Polonoroeste after 1987, Washington, D.C., March 3, 1991 and August 4, 1993.

37. Interviews with Carvalho Filho and Mindlin and The World Bank 1991, 71.

38. Interview with Mindlin.

39. Ibid.

40. Interview with Carvalho Filho.

41. Ibid.

42. "Statement of Bruce Rich on behalf of Environmental Defense Fund and National Wildlife Federation concerning the Environmental Performance of the Multilateral Development Banks and the International Monetary Fund, before the Subcommittee on Foreign Operations Committee on Appropriations United States House of Representatives, April 24, 1989" (p. 1). The Sierra Club, another environmental organization supporting the MDB Campaign, had approximately 496,000 members in 1989 (Mitchell 1990, 92).

43. Interview with Carvalho Filho.

44. Interview with Mindlin.

45. Browder 1987.

46. The World Bank 1991, 67.

47. Mindlin 1988; and *"ITC embarga desmatamento da FUNAI," Correio de Noticias,* February 2, 1985.

48. CEDI and CONAGE 1986.

49. For details, see Pinheiro and Leão 1989; The World Bank 1992c; Freitas and Soares 1994.

50. IBGE 1993, 2–27.

51. Little research and a rushed project design overestimated the farming potential of most of Rondônia's soils.

52. The use of different methodologies to estimate deforestation in Amazonia in general, and in Rondônia in particular, have generated conflicting results. Yet for illustrative purposes, I present some of the available statistics in tables 3.1 and 3.2.

53. For a detailed study on the relationship between human occupation of Rondônia and deforestation, see Fearnside 1989.

54. Goodland 1985; Carvalho Filho 1987.

55. See Carvalho Filho 1987. Until 1986, Polonoroeste had settled 4,600 families, which represented only 30 percent of its initial goal (p. 10). Also, The World Bank (1992c) evaluation of the project's small-farmers settlement component indicates a

TABLE 3.1
Landsat Surveys of Forest Clearing in Brazil's Amazonia

FU	Area (1,000 Km²)	Area Cleared (1,000 Km²) By				% of State Territory Classified as Cleared By			
		1975	1978	1980	1988	1975	1978	1980	1988
RO	243	1,216	4,184	7,579	58,000	0.3	1.7	3.1	23.7
MT	881	10,124	28,355	53,299	208,000	1.1	3.2	6.1	23.6

SOURCE: Adapted from Pearce and Mayers 1990, 386.

FU: Federal Unit; RO: Rondônia; MT: Mato Grosso

TABLE 3.2
Legal Amazonia—Extent of Gross Deforestation (Km^2)

FU	1978	1988	1989	1990	1991
RO	4,200	30,000	31,800	33,500	34,600
MT	20,000	71,500	79,600	83,600	86,500

SOURCE: Adapted from INPE 1992.

FU: Federal Unit; RO: Rondônia; MT: Mato Grosso

high settlers' turnover rates. A case study of one of Polonoroeste's settlement projects showed that 24 percent of all settlers interviewed in 1985 had moved on one year later (p. 81).

56. Bunker 1985; Mahar 1989; Hecht 1983 and 1985; among others.

57. Binswanger 1990; IBASE 1985; Millikan 1992; among others. Carvalho Filho (1987) provides some numbers on the expansion of cattle ranching in Rondônia: between 1980–1985, the number of cattle grew threefold. The annual expansion in the number of animals in the state was 25 percent, as opposed to 1.5 percent in Brazil (p. 18).

58. The World Bank 1992c; interviews with Avelar and Carvalho Filho.

59. Leonel 1991; Brunelli 1986.

60. Rich 1990; LePrestre 1989; Keck and Sikkink 1998; among others.

61. Interviews with World Bank staff and members of the Polonoroeste management team Daniel Goss, John Redwood III, and Luis Coirollo (former task manager), Washington, D.C., respectively March 8, 12, and 3, 1991.

62. Several of my interviewees (activists, consultants, and World Bank officials) stressed the importance of the World Bank's role in pressuring the Brazilian government to demarcate indigenous lands throughout Brazil in the 1980s and 1990s. Without questioning such an assessment, data show that, in the specific case of the region affected by the Polonoroeste project (the states of Rondônia and Mato Grosso), the World Bank's role was limited. Of the thirty-two indigenous areas in Rondônia and Mato Grosso, twenty-eight were included in the Polonoroeste Amerindian Special Project (ASP). Of these twenty-eight, eighteen had been demarcated before the project started. Their demarcation was, therefore, outside the scope of the MDB/World Bank pressures; eight indigenous areas were still not demarcated as of 1987. Finally, six were demarcated within the scope of the Polonoroeste project. In other words, during the 1980s, the MDB Campaign and the World Bank, at best, may have influenced outcomes in less than 20 percent of indigenous areas in the Polonoroeste region (information on the status of demarcations was found in the following documents: Instituto Brasileiro de Geografia e Estatistica (IBGE) (1993), *Recursos Naturais e Meio Ambi-*

ente—Uma Visão do Brasil, IBGE, Rio de Janeiro; The World Bank (1991), "Project Completion Report Brazil—Northwest Region Integrated Program—Phase I Agricultural Development and Environmental Protection Project (Loan 2060–BR)," and The World Bank, Washington, D.C., April; The World Bank (1992), Staff Appraisal Report—Brazil Rondônia Natural Resources Management Project, February 27, The World Bank, Washington, D.C.

CHAPTER 4. "LOCALIZING" TRANSNATIONAL ACTIVISM—
SUCCESS AND FAILURE

1. Interviews with Ana Avelar, president of the Rondonian NGO, INDIA, and former public employee of the Rondônia State, Rondônia, November 24, 1994, and with Maria Emilia da Silva, subsecretary for Planafloro monitoring, and former public employee of the Rondônia State, Rondônia, November 24, 1994.

2. IAMA's partner was *Ação Ecológica Vale do Guaporé* (ECOPORE), a one-staff ecological NGO headquartered outside Rondônia's political center. Currently, ECOPORE has an office in Rondônia's capital and has become the most visible environmental NGO in the state.

3. The others being Amerindians, small farmers, and small communities along Rondônia's main rivers.

4. One example is Ana Avelar, who worked for both IAMA and IEA and later became the president of INDIA, a local environmental organization.

5. Keck 1993.

6. The environmental management component of Planafloro accounted for less than one-third of project funds (US$64.8 million), while the agro-forestry and socioeconomic infrastructure components received, respectively, US$81.4 and 71.5 million (The World Bank 1992a, 76).

7. For details on state politics surrounding the preparation and signing of Planafloro, see Keck 1998.

8. An extensive discussion of the different approaches to the concept of environmental sustainable development is provided in the introduction. Briefly, an actor's approach to the concept includes both its theoretical/philosophical understanding on how to harmonize the ideals of material improvement and conservation of natural resources, and its ideological, political, and material interests in these processes. As discussed in the introduction, "conservative" and "mainstream" approaches to environmentally sustainable development distinguish between socioeconomic and political structures and efforts toward environmental sustainability. Alternative approaches, in this book labeled socioenvironmental development, *integrate* environmental, economic, social and political concerns.

9. With the help of IEA's network in the United States, Mendes gathered international support for the implementation of rubber tappers' extractive reserves in Ama-

zonia. Since 1985, with the creation of the National Council of Rubber Tappers (CNS) in Brazil, it has become a major demand of groups concerned with sustainable economic alternatives in tropical forests.

10. Mary Alegrette, head of IEA, first conceived the idea of extractive reserves in partnership with the rubber tappers' council. The idea quickly received the endorsement of the international environmental community as a potential way of extracting economic value from the forest without destroying it.

11. Francisco Alves Mendes Filho, letter to World Bank president Barber Conable, October 13, 1988; and Rich 1994, 167.

12. Chico Mendes's assassins were quickly identified as hired hands of a local cattle rancher who had made several public threats against Mendes's life. Despite an investigation and trial riddled with corruption, pressures from the national and international communities have guaranteed that Mendes's assassins remain in jail to this day.

13. Keck 1993.

14. For details on the cleavages between Rondônia's and Acre's rubber tappers see Keck 1993.

15. Keck 1998.

16. EDF letter to the World Bank U.S. executive director, Patrick Coady, January 9, 1990, and Rich 1994.

17. José Lutzenberger's letter to the World Bank president, Barber Conable, March 22, 1990.

18. Local Rondonian NGOs signed a formal agreement *(Protocolo de Entendimento)* with the state in which the latter committed to the protection of extractive reserves and indigenous areas; the prevention of land titling and settlement initiatives, logging, and mining in these areas; the creation of a support program for rubber tapper communities; the elimination of deforestation as a criterion for determining land titling; and the commitment to strengthening Environmental Impact Assessments for productive activities in the state *(Protocolo de Entendimento,* item 8).

19. Most international environmental NGOs and significant sectors of the international environmental community (specialized media and selected development agencies) fully embraced the social and economic dimensions of environmental issues during the UNCED 1992. Southern environmental groups participating in the Parallel Event to the Conference played a key role in shaping this process.

20. Interview with Ana Avelar.

21. Namely, Brent Millikan, a graduate student from Berkeley. Millikan had been in Rondônia since the 1980s and had participated in the Rondônia network's initiatives against Polonoroeste. In the 1990s, Millikan continued to be a key asset for the network, becoming "the source for the lion's share of the technical and institutional research that has been produced [on Planafloro]" (Keck 1993, 21), and providing leadership for many of the strategies deployed by the network in Rondônia and abroad.

22. Keck (1998) argues that local civil society groups did not realize that participation in the CD was a token gesture from the government. It soon became evident that major decisions on Planafloro implementation occurred outside the scope of the CD and were essentially determined by the pork barrel practices that characterize Rondonian politics. My evaluation is that rather than being naïve about Rondonian politics, NGOs pragmatically occupied the space that had become available to them within Planafloro, hoping, at least, to gain more information and insight into the project.

23. Rondonian NGOs and the Rondonian government: *"Protocolo de Entendimento,"* June 20, 1991, and IEA: "Governmental Policies, Environmental Degradation, and the Challenge in the Amazon Basin: the Case of Rondônia, Brazil (1992), Project Proposal submitted to WWF-Brazil (mimeo).

24. The approval of a socioeconomic and environmental zoning law for the state of Rondônia was a crucial precondition of Planafloro. The law, once passed, would make it illegal for settlers, governmental agencies, and the private sector to exert activities in any given area that did not agree with that area's socioenvironmental vocation. The Rondonian Legislative Assembly eventually approved the zoning law, which was based on initial studies of soil and natural resources in Rondônia ("first approximation"). The approval came shortly before Planafloro was signed, and after much opposition from political groups linked to mining and logging interests. The law divided the state into several areas, classified in six zones. Zones 1 and 2 identified areas adequate for agriculture and livestock activities, as well as intense settlement. Zones 3 to 6 ranged from areas where only limited extractive activities should be allowed, to areas where human occupation should be entirely prevented (conservation units). One key goal of Planafloro was to conduct more detailed zoning studies ("second approximation") as the basis for an updated zoning law for the state.

25. Interview with Ana Avelar.

26. Rondônia NGOs Forum, *"Carta Aberta aos Diretores Executivos do Banco Mundial,"* Porto Velho, March 12, 1992.

27. It is interesting to note that some government bureaucrats in Rondônia perceived the World Bank as "a spokesperson for the environmental entities . . ." (interview with Pedro Wilson, forest engineer in the Rondonian Secretariat of Agriculture, November 1994).

28. Namely, those by the MDB campaigners and by other transnational activists who have pressured the bank to redefine its development role.

29. Interview with Antenor Karitiana, leader of the Karitiana Indigenous peoples, Rondônia, November 25, 1994.

30. The history of the National Institute for Colonization and Agrarian Reform in Amazonia is complex. As the agency in charge of settlement, land titling, and agrarian reform policies, the institute is a major electoral asset anywhere in Brazil, and in Amazonia in particular. In Rondônia, political elites and landowners' oligarchies are among the groups that have benefited the most from the agency's policies. Its process of legalization of squatters' rights is often manipulated for electoral purposes. This partially explains the agency's resistance to the Rondônia zoning law. In the politics of the

state, recognition of squatters' rights is a far more popular measure than their eviction based on the law. Despite the merits of the zoning initiative, its original methodology neglected social analyses that could have highlighted this issue. Zoning studies prioritized technical aspects of soil conditions and resources potential, disregarding the complexity of situations in which human communities had already been established in areas destined for strict conservation.

The policies of the colonization agency, however, cannot be explained only by a populist rationale. Corruption and embezzlement have been frequent charges involving some of the agency's expropriation processes. The agency has considered expropriating large estates in areas inadequate for agricultural activities (zones 3 to 6 according to the 1991 zoning law). Yet such expropriations have only entailed further corruption. Price overvaluation is common in these cases. The original landowner benefits twice from the expropriation process: he/she has a guaranteed buyer for a property that, because of the zoning law, has lost much of its economic value, while also obtaining a higher-than-market price for the property.

Often, justification for an overvalued price is based on "improvements" made on the estate to be expropriated. Internal regulations of the colonization agency still consider deforestation (usually for the establishment of pastures) as an indication of "productive intentions," and, therefore, as an "improvement." This practice remains legal in Rondônia, despite the fact that one precondition of Planafloro was that INCRA and the Rondonian government establish an agreement "concerning land regularization policies and practices to be made consistent with the objectives of sound forest protection and management" (The World Bank 1992a, 44).

The logging industry has often benefited from the colonization agency's disregard for zoning regulations and environmental impact assessments for its settlement projects in Rondônia. Logging groups have actively encouraged what is referred to in Brazil as the "invasions industry." This "industry" is based on mutual advantages for those involved in it. Illegal logging activities have led to the opening of roads, thus facilitating the access of migrants to protected areas. In turn, the presence of squatters in these areas has led to political and legal dilemmas, which are usually resolved by the state's decision to downsize conservation units, thus allowing INCRA to distribute land rights to invaders. The logging industry thus benefits from both the by-product of peasants' clearings (wood) and access to larger portions of unprotected forest.

Finally, the ability of the colonization agency to disregard zoning regulations and resist pressures by environmental activists comes from two sources. First, until 1995, the agency had control over the vast majority of public lands in Rondônia. Despite Rondônia's change of status from a federal territory to a state in 1982, it did not gain full ownership of its public lands. Planafloro specifically addressed this issue by requesting, as a loan precondition, the creation of a state land institute and the transferring of the functions (and landownership) from the federal colonization agency to the new institute. The purpose of this precondition was to make land issues subordinate to state policies (namely, the zoning law). INCRA, however, refused for many years to transfer public lands to the Rondonian state. The second explanation for the autonomy of the colonization agency relates to its association, at the local level, with highly influential politicians. In fact, leaders of Rondonian political parties usually are the ones who appoint the agency's regional superintendent. This has lent significant autonomy to INCRA's local chapter vis-à-vis its federal office in Brasília. Formal commitments made by the latter in response to pressures by the World Bank and environmental activists rarely inhibit the actions of the agency's local bureaucracy.

31. World Bank letter to the Brazilian minister of economy and planning, Marcílio M. Moreira, June 22, 1992; letters by the Rondônia Forum to the president of the Republic of Brazil, May 29, 1992, to the Ministry of Economy and Planning, May 29, 1992, and to INCRA's president, May 29, 1992; August 14, 1992; September 8, 1992; February 28, 1993; and April 15, 1993.

32. Although some areas in Rondônia are considered adequate for agriculture, they have suffered an intense process of land concentration. The colonization agency has resisted any form of land redistribution scheme in these areas that would benefit small farmers.

33. Rondônia NGOs Forum: *"ONGs de Rondônia Criticam Politica Fundiaria do INCRA,"* (a call for action), Porto Velho, April 20, 1993.

34. Interview with Hélio Madalena, legal advisor for the Rondônia Forum, Rondônia, November 24, 1994.

35. Letter from the Forum of NGOs and Social Movements of Rondônia to Mark Wilson, Division Chief, Natural Resources and Rural Poverty, Latin American and the Caribbean Region, World Bank, June 15, 1994.

36. Millikan 1995.

37. Interview with Stephen Schwartzman, EDF's senior environmentalist, March 1, 2001. Sectors of the Rondônia media also interpreted the August 1994 World Bank mission as a preparation step toward the suspension of disbursements (*O Progresso,* August 5, 1994, Rolim de Moura, Rondônia).

38. Respectively, interviews with José Maria dos Santos, May 23, 2000, and José Carlos Gadelha, at the time coordinator of the Rondônia's chapter of the Pastoral Land Commission (CPT), Rondônia, November 22, 1994.

39. Interview with Luis Rodrigues de Oliveira, Rondônia Forum's executive secretary between 1994 and 1996, Brasília, March 21, 1995.

40. Ibid.

41. Instituto de Estudos Amazônicos e Ambientais (IEA) (1992): *"Breve Diagnóstico Institucional de Órgãos Executores do Plano Agropecuário e Florestal de Rondônia (Planafloro), Relatório Preliminar,"* November, Porto Velho (mimeo).

42. Interview with Iremar Ferreira, staff of the Indigenous Peoples' Missionary Council, Rondônia, November 25, 1994.

43. Namely, the Center for Indigenous Rights *(Núcleo de Direitos Indígenas),* whose cadres later joined the Social Environmental Institute (*Instituto Socio Ambiental*—ISA).

44. Some criticisms of the project design refer to the excessive number of bureaucracies that it involved and its lack of adequate provisions to guarantee the institutional capacity of implementing agencies.

45. Interview with Gabriel Ferreira, consultant for the United Nations Development Program, UNDP, Rondônia, November 23, 1994.

46. Interview with Ana Avelar.

47. Interview with José M. dos Santos, at the time OSR vice president, Rondônia, November 22, 1994.

48. Interview with Antenor Karitiana.

49. Interview with José Maria dos Santos and OSR: *"Extrativismo em Rondônia,"* (a call for action), December 1994.

50. *"Indios tentam invasão"* and *"Indios ameaçam operação"* articles in the Rondonian newspaper *Diário da Amazônia*, Rondônia, September 28, 1994, and *"Indios tentam tomar FUNAI de Rondônia," Folha de São Paulo,* São Paulo, September 30, 1994.

51. Most Rondonian activists whom I interviewed in November 1994 addressed this issue.

52. In this specific context, citizenship rights refer to the right to participate in and have one's interests addressed by development policies for the state.

53. Interviews with Ana Avelar and Aurélio Vianna, executive secretary for the *Rede Brasil Para Instituições Financeiras Multilaterais* (Brazil's Network) (1995–1998), Brasília, May 19, 2000.

54. Interviews with José Gadelha and Marcelo de Pádua, Planafloro geologist in the Department for Special Projects in Brazil's Ministry of Regional Integration, Brasília, September 26, 1994.

55. Interviews with Hélio Madalena, Mauro Leonel, Planafloro's consultant and IAMA's director, São Paulo, September 12, 1994; and Pedro Wilson, staff, Secretariat of Agriculture, Porto Velho, November 25, 1994.

Chapter 5. Listening to the Grassroots—
The Rondônia Network and Local Politics

1. Udall 1998, 414. On the Inspection Panel, see also Fox 2000, and D. Clark, J. Fox, and K. Treakle 2003 (forthcoming).

2. Interviews with Francisco Vita, World Bank task manager for Planafloro, Mato Grosso, June 9, 2000, and John Garrison, World Bank—NGO liaison with civil society, Brasília, May 19, 2000.

3. Interview with Luiz Oliveira, Rondônia Forum's executive secretary between 1994 and 1996, Brasília, March 21, 1995.

4. Here the role of international NGOs in the Planafloro inspection panel claim must be stressed. When I met Brent Millikan, one of the main authors of the eighty-one-page claim document in Brasília in March 1995, he was already working on a background document for the claim with Roberto Smeraldi (FoE). Yet the represen-

tative of the forum, with whom I spoke on the same occasion, stated that, at that point, there was no formal involvement of the forum in the process beyond its members' awareness of Millikan and Smeraldi's work.

5. Interviews with Roberto Smeraldi, coordinator of Friends of the Earth—Amazônia Program, São Paulo, June 5, 2000, Oliveira, and Jean Pierre Leroy, consultant for the Federation of Agencies for Social and Educational Assistance (FASE), Rio de Janeiro, December 11, 2000.

6. Interview with Smeraldi.

7. Interviews with Garrison, Vita, Smeraldi, and Oliveira, and with Aurelio Vianna, former executive secretary for the Brazil Network on Multilateral Development Banks—Rede Brasil, Brasília, May 2000.

8. Interview with Vianna.

9. Interviews with José Maria dos Santos, former vice president and current president of the Organization of Rondonian Rubber-Tappers and with Anselmo Abreu, president of the Federation of Agricultural Workers of Rondônia, Rondônia, May 23, 2000.

10. Forum das ONGs e Movimentos Sociais que Atuam em Rondônia and Friends of the Earth/Amigos da Terra Programa Amazonia 1995, 10.

11. The request for inspection also denounced the World Bank's failure to oversee Planafloro's administration and its investment and media programs.

12. Forum das ONGs e Movimentos Sociais que Atuam em Rondônia and Friends of the Earth/Amigos da Terra Programa Amazonia 1995, 5.

13. David Hunter, "The Planafloro Claim: Lessons from the Second World Bank Inspection Panel Claim," *www.ciel.org/planfl2*, Millikan 2001, and interview with Oliveira.

14. Millikan 2001 and interview with Brent Millikan, former consultant for the Rondônia Forum, Brasília, May 19, 2000.

15. *www.ciel.org/planafl2.html*, p. 7

16. The Rondonian groups were represented by the forum's executive secretary, by the leader of the rubber-tappers, and by the leader of an Amerindian group.

17. Interview with Oliveira.

18. Between 1994 and 1999 the panel received fifteen requests for investigation. Of these, the panel concluded that three were not eligible for investigation, and five did not warrant an investigation. Of the seven remaining, one is still being evaluated (China: Western Poverty Reduction Project). The panel recommended the investigation of six projects, but the board of directors only approved two of them. Four requests were "resolved" by the board's accepting action plans rather than full panel investigations.

19. For details see *www.ciel.org/planafl2.html*. In fact, new language was included in the "Conclusions of the Board's Second Review of the Inspection Panel," (item 14). See *www.worldbank.org/html/extdr/ipwg/secondreview.htm*.

20. "Cronograma dos Fatos," *Noticias do Forum* (newsletter) 4, year 3 (December 1995).

21. "Banco Mundial tenta evitar a denúncia," *Noticias do Forum* 4, year 3 (December 1995).

22. Smeraldi and Millikan 1997.

23. Keck 1998, Millikan 2001, and Hunter.

24. Interviews with Stephen Schwartzman, senior environmentalist at the Environmental Defense, Washington, D.C., March 1, 2001, and Vita.

25. Interviews with Vita and Garrison.

26. Interview with Garrison.

27. Ibid.

28. Jesus 1998 and interview with Luiz Oliveira.

29. Interviews with Millikan, Iremar Ferreira, consultant for CUNPIR, Rondônia, May 23, 2000, and José Gadelha, former executive secretary of the Rondônia Forum, Rondônia, May 23, 2000.

30. The one real innovation of the Planafloro's restructuring process is the PAIC. All other components were simply rearranged and renamed, with minimal budget changes. The PAIC received US$ 20,000,000 (twenty million dollars, which represents 12 percent of the World Bank's total funding for Planafloro). This amount, however, grows in significance when placed in the context of the 1996 negotiations. At the time, the main issue was not to reallocate Planafloro's entire budget, but to try to make the best possible use of its remaining funds (US$72,000,000). Under this perspective, the PAIC meant that the Rondonian grassroots groups gained direct access to 28 percent of Planafloro's remaining funds.

31. I have inferred the existence of an alternative approach to environmentally sustainable development and to the ways by which it could be made operational within the restructured Planafloro project from interviews with representatives of Rondonian civil society organizations in May 2000, and from the analysis of documents on the Program for the Support of Community Initiatives.

32. The community initiatives program provides grants of up approximately US$80,000. The beneficiary community must provide 20 percent of counterpart funds for productive, social, and infrastructure activities, but in the case of proposals that are exclusively environmental, counterpart requirements decrease to 10 percent.

33. SEPLAN/Planafloro 1997, 5–7, 13–14 and Oliveira 1998.

34. According to Renato da Costa Melo, manager of the Community Initiatives Program, the composition of the council was, in 2000, as follows: government seats were filled by representatives of the secretariats of agriculture, environment, education, health, tourism, and planning, and of the agency for technical assistance and rural development, EMATER; civil society seats were filled by officials from private sector organizations (the Federation of Agricultural Producers of Rondônia and the Federation of Industries of Rondônia), and from NGOs and grassroots organizations, namely,

the Landless Peoples' Movement, the Organization of Rondonian Rubber Tappers, the Coordination of the Union of Indigenous Peoples and Nations of Rondônia, the Center for Amazonian Research at Cunian, and the Federation of Agricultural Workers of Rondônia. Interview, Rondônia, May 23, 2000.

35. Once a project is approved the government has to establish a joint account between the representative of the awarded group (NGO, cooperative, Amerindian community) and the Planafloro's executive secretary. The intent of the joint account is to increase the autonomy of beneficiary communities and diminish red-tape over financial transactions.

36. Interviews with Melo, Smeraldi, and Luiz C. Pinajé, staff, World Wildlife Fund—Brazil, Brasília, May 26, 2000.

37. For details, see Millikan 1998 and 2001.

38. Interview with Santos, and phone conversation, May 17, 2001.

39. Interviews with Oliveira, Abreu, Santos, Gadelha, and Ferreira.

40. Interview with Iremar Ferreira.

41. Ibid.

42. Interview with Anselmo de Abreu.

43. Browder 1998, 14.

44. Interview with Abreu. The president of FETAGRO establishes a clear causal relation between Planafloro/PAIC and the opportunities for civil society participation generated by these initiatives, and FETAGRO's increased environmental awareness. This is corroborated by Pinajé, from the WWF-Brazil, who singled out FETAGRO as one of the organizations with the highest level of environmental consciousness within Brazil's Confederation of Agricultural Workers (*Confederação dos Trabalhadores na Agricultura*—CONTAG).

45. Interview with Abreu.

46. "Forum das ONGs e Fetagro realizam encontro sobre Planafloro," *Notícias do Forum* 34, year 7 (October 1999); and "Forum das ONGs e Fetagro cobram recursos dos PAICs" (a report on the Mobilization Day/Action for Citizenship of the PAIC beneficiaries), *Notícias do Forum* 35, year 7 (November/December 1999).

47. The term second approximation refers to more detailed satellite images and mapping efforts than those conducted at the first round of studies about Rondonian soils.

48. Interview with Santos. Difficulties were ameliorated after the Rondônia Forum started to conduct educational workshops that explained the zoning process.

49. Interviews with Santos, Abreu, and Ferreira.

50. Interview with Garrison.

51. "Um Novo Planafloro," in *Notícias do Forum* 8, year 4 (July 1996), 1.

52. Interview with Gadelha.

53. Interview with Ferreira.

54. Interview with Santos.

55. Browder 1998. The evaluator proposes revised guidelines for the approval of community projects. These should address the fact that many projects lack functional integration among their different components. In additions, guidelines should require increased levels of partnership between recipient organizations and other institutions, particular local governments, thus avoiding the possibility that community initiative projects funded by Planafloro become substitutes for public investments.

56. Interview with Iremar Ferreira.

57. Interview with José Gadelha.

58. Interview with Brent Millikan.

59. Governor Raupp's control of program's funds was in great part granted by World Bank requirements on disbursements: funds can only be transferred to state accounts. NGOs formulating the community initiatives program attempted to guarantee the direct access of beneficiary organizations to funds. The solution, as mentioned above, was to create joint bank accounts between the state and each of the program's recipient organizations. By the end of 1998, many organizations had already received notice that funds had been deposited in their accounts and were surprised by the state's withdrawal of these funds without previous warning.

60. Interview with Iremar Ferreira.

61. Interview with José Maria dos Santos.

62. Interview with José Carlos Gadelha.

63. Interview with Luis Carlos Pinajé.

64. Fox 2000b.

Chapter 6. Environmental Activism beyond Brazil I— The Struggle against Oil Exploitation in Ecuador

1. It is beyond the scope of this chapter to discuss the economic advantages of oil exploitation for Ecuador. Suffice it to say that there is an ongoing discussion in academic and political circles about the extent to which oil exports have contributed to improve Ecuador's economic standing or whether they have increased social and economic inequalities in the country and contributed to heighten its foreign debt. Another contentious issue not addressed here is the extent to which oil exploration is a more environmentally friendly alternative than other economic activities in the rainforest.

2. Kimerling 1991 and CESR, Oxfam America, and Rainforest Action Network 1996.

3. See Treakle 1998.

4. See Selverston-Scher 2001.

5. The Huaorani are an indigenous group in the *Oriente* that in the 1980s were still in very early stages of contact with Ecuadorian society.

6. For details see Kane 1996, Kennedy Jr. and Scher (n/d), and the debate between Kane and Kennedy in *The New Yorker* (September and October issues, 1993), and Jochnick and Garzon 2001.

7. Jezic 2001.

8. CESR, Oxfam American, and RAN.

9. Selverston 1997, Collins 2000, and Selverston-Scher 2001.

10. Selverston-Scher 2001, 29.

11. Selverston 1997, 179.

12. Selverston-Scher (2001) argues that the 1990 uprising had, besides structural causes, such as land conflicts, unemployment, and economic recession, ideological causes as well, mainly related to issues of indigenous identity (p. 61).

13. Treakle 1998, 245.

14. Pacari 1996 and Treakle 1998.

15. Nina Pacari, congressional representative for Pachakutik, cited in Collins 2000, 46.

16. The formal name of the Amazon Coalition is Coalition for Amazonian Peoples and their Environment. In 1999 it became the Amazon Alliance.

17. See for instance, Kane 1996, Selverston-Scher 2000, and Jezic 2001.

18. Interview with Selverston-Scher, director of the Amazon Coalition between 1991 and 1999, Washington D.C., March 12, 1997, and Selverston-Scher 2000.

19. Benavides 1996.

20. Jezic 2001, 186.

21. Interview with Selverston-Scher.

22. Benavides 1996.

23. "The projects of colonization in the Amazon and the coastal regions have not been favorable for our peoples; the situation has worsened. Through the system of colonization, the theft of natural resources has been legitimized, as well as the appropriation of natural resources by large businesses." Luis Macas, former president of CONAIE, cited in Ortiz-T 1997.

24. Selverston-Scher 2000, 4.

25. Selverston-Scher 2000 and informal conversations with Cathy Ross, coordinator for Oxfam America's Amazon Program, April 1996.

26. Jezic 2001.

27. Jezic 2001 and Jochnick and Garzon 2001.

28. In hindsight, their most important consequence was raising awareness among other local groups affected by oil operations, as will become evident in the discussion of the campaign against ARCO.

29. *Maria Aguinda et al. v. Texaco*, filed on November 3, 1993, in the U.S. District Court for the Southern District of New York.

30. Jezic 2001, 185.

31. Jezic 2001, several issues of the newsletter of the committee of Texaco plaintiffs *(Comite de Demandantes contra la Texaco)*, and *www.texacorainforest.com*, among others.

32. While the coalition's work was significant for the legitimacy of international pressures in the matter, it was the domestic activism of members of the anti-oil network that determined the government's change of position. For details, see Jezic 2001.

33. The responses of the Ecuadorian government to the lawsuit against Texaco are best summarized as contradictory. Different administrations and different agencies (sometimes within the same administration), have at times supported, and at times repudiated the lawsuit. In an affidavit to the U.S. court signed in 1994, the then minister of Energy and Mines, Manuel Navarro, recognized that the poor, indigenous peoples and marginalized sectors in Ecuador have difficulties in obtaining justice within the country. Yet the office of President Sixto Durán Ballén (1992–1996), systematically opposed the lawsuit under the justification that it threatened Ecuador's national sovereignty. Eventually, the opposition of the Ballén administration provided the justification for the dismissal of the case in 1996. Such a decision was reversed almost immediately, as a result of the efforts of the anti-oil network. The network succeeded in obtaining a letter from the incoming administration of president Abdalá Bucaram indicating that the lawsuit did not harm the country's national sovereignty. This position was reiterated by president Fabián Alarcón, who replaced Ballén following his impeachment in 1997. The Alarcón administration, under intense lobbying by the anti-oil network, went a step farther and declared its support for the lawsuit. Such a support was later withdrawn during the Mahuad administration, elected in 1998. Nevertheless, the anti-oil network was still able to guarantee the cooperation of attorney general Ramón Jiménez Carbo during that period. (This summary is based primarily on information found in Jezic 2001.)

34. Selverston-Scher 2000.

35. Press release from the Coalition for Amazonian Peoples and their Environment, November 25, 1998, "Indians Sue Texaco for Environmental Destruction—Ecuador Jeopardizes Case," Letter from the Coalition for Amazonian Peoples and their Environment to Jamil Mahuad, president of Ecuador, November 25, 1998, and *Comite de Demandantes contra la Texaco*, "La Texacontaminacion en el Ecuador, February, 1999" (newsletter).

36. Amazon Coalition's call for registration, "Amazon Coalition-COICA Workshop Regional Strategies in Defense of the Amazon Against Oil Development," November 9, 1995, and Jezic 2001.

37. Interview with José Maria Sadaba, Capuchin priest working in the city of Coca *(Oriente)*, Quito, Ecuador, August 25, 2001.

38. Newsletter, "Voz de la CONFENIAE" 18, year 3 (October-November-December 1995), 6–7, and correspondence between CONFENIAE and *Acción Ecologica*, October 1995, and between CONFENIAE and Jim Ford, manager of the Santa Fe Oil Company, May 10, 1995.

39. Interview with Alexandra Almeida, *Acción Ecologica*, Quito, Ecuador, August 29, 2001.

40. Interview with José Sadaba.

41. Jezic 2001.

42. Jochnick and Garzon 2001.

43. Interview with Alexandra Almeida. Almeida points out that the episode had repercussion for the legitimacy of the Ecuadorian indigenous movement as a whole. Many of the individuals implicated had been elected as legislative representatives and lost their mandates.

44. *Comite de Demandantes contra la Texaco*, "La Texacontaminacion en el Ecuador, February, 1999" (newsletter).

45. Letter from Luis Yanza, President of the *Frente de Defensa de la Amazonia*, to Ivonne Ramos, *Accion Ecologica*, December 16, 1999.

46. Since 1988, OPIP has had an institutional presence in the area. Before the official start of the campaign, it had already demanded negotiations with company executives. At times, it even detained oil workers for several days to force a dialogue (for details see Mendez et.al 1998).

47. McCreary 1995.

48. For instance, failure to conduct baseline environmental studies before the start of seismic investigations, lack of environmental guidelines available to contractors, the conditions of abandoned well sites, decrease in wild game in parts of the region, and unsuccessful re-vegetation efforts, among others (Mendez et al. 1998). For details on the Berkeley review see McCreary 1995.

49. Interview with Alexandra Almeida; interview with Alfredo Luna, lawyer with CORDAVI *(Corporacion de Defensa de la Vida)*, Quito, Ecuador, September 28, 2001.

50. Quichua leader Leonardo Viteri, quoted in The Seventh Generation Fund, "The Struggle of the organization of indigenous Peoples of Pastaza, OPIP," January 1994 (mimeo).

51. Sawyer 1996.

52. Notes on the meeting between Luiz Macas, former president of CONAIE, with staff from several Washington, D.C.-based NGOs, on September 21, 1994.

53. Mendez et al. 1998.

54. Sawyer 1998 and "OPIP y petrolera ARCO buscan un acuerdo," *Hoy* (Ecuadorian newspaper), December 20, 1993. ARCO's response to the charges is that

it did not fully comprehend the politics of indigenous peoples' associations in the Ecuadorian Amazon.

55. Sawyer 1996.

56. Mendez et al. 1998.

57. Sawyer 1998.

58. Bebbington 1998 and "Arco Evacua sus Trabajadores Mientras los Sequestros Continuan," July 29, 1998 *www.unii.net/confeniae/espanol/noticias/arco.*

59. The primary spokesperson for network members in negotiations against oil activities in the *Oriente* and in Block 24, in particular, has become the Ecuadorian government. Network members understand that the national government should be responsible for guaranteeing that oil companies respect the country's environmental and social legislation as well as the specific terms of their concession agreements. This approach has gained strategic importance in a moment when ARCO has begun to transfer its business in Ecuador to other oil companies such as Agir and Burlington.

60. Interviews with Alexandra Almeida and Alfredo Luna.

61. Jezic and Jochnick 2000, "The Meaning of a Legal Victory in the Ecuadorian Amazon," *www.cceia.org/hrdspring2000.*

62. Jezic and Jochnick 2000, and FIPSE and *Centro de Derechos Economicos y Sociales* (CDES): *"Victoria en una Accion de Amparo en contra de Arco Oriente Inc"* (newsletter, n/d).

63. Interview with Alfredo Luna.

64. Keck and Sikkink 1998.

65. Fox and Brown 1998.

66. Treakle 1998, 251.

67. Interviews with Alexandra Almeida and Alfredo Luna.

68. Interview with José Sadaba.

69. Since 1997 the New York–based Center for Economic and Social Rights has established an office in Quito.

70. Interview with Alexandra Almeida. The history of the Cofan struggle, however, is atypical in the context of Ecuadorian local groups. For details, see Tidwell 1996.

CHAPTER 7. ENVIRONMENTAL ACTIVISM BEYOND BRAZIL II— THE STRUGGLE AGAINST LARGE DAMS IN INDIA

1. The reader should be advised that this chapter does not provide new information on the struggle against the Sardar Sarovar dam and the international mobilization it has triggered. There is extensive bibliography on the issue (much of it is

listed in this book's Bibliography). Updated information can also be found in the web pages of the International Rivers Network (IRN) *www.irn.org* and of the Friends of the River Narmada *www.narmada.org*. What this chapter offers—which is unprecedented to my knowledge—is an analysis of the network efforts "from within" as well as a focus on how local groups have affected network strategies and, conversely, how they have been impacted by such strategies.

2. Roy (1999) thus refers to large dams, making an ironic reference to Nehru's statement, "Dams are the Temples of Modern India," which he came to regret.

3. The controversy can be explained, in part, by different methodologies and definitions of affected people. Some consider only the population displaced by the reservoir, other include those affected by the construction of the canal and irrigation network. Still others count individuals that will not be displaced but will suffer social and economic consequences of the alterations produced by the project.

4. Fisher 1995, 13.

5. Medha Patkar's interview with WBEZ Radio, Chicago, November 26, 1999.

6. Roy 1999, 42.

7. See Patel 1995, Jay Sen, " Milestone or Landmark? The Narmada Case in Historical Perspective," published in *Mainstream*, March 11, 2000, and available at *www.narmada.org./articles/JAI_SEN/milestone*.

8. Adivasis were once referred to as tribal people. They live mostly in Indian forests, and in the particular case of the Sardar Sarovar project area, their villages are located on the hills of the Nimar valley, most of which will be flooded by the SSP reservoir. In the Indian caste system, they are considered scheduled (lower) castes.

9. For a detailed account on the Sangath's role in organizing Adivasis, see Baviskar 1995.

10. Baviskar 1995.

11. Arch-Vahini's position changed in June 1995 when it resigned from the Government Committee on Resettlement and circulated a letter admitting enormous problems with the SSP resettlement program (Namada Bachao Andolan: "The Narmada Struggle—A Brief Chronology since 1993" [October 1995]).

12. Sen 2000.

13. Smitu Kothari, "Who can speak for the people?" *www.narmada.org/debates/ramguha/smitusresponse*.

14. Patel 1995.

15. Baviskar 1995.

16. NBA 1990, cited in Baviskar 1995, 206.

17. For details see Morse and Berger 1995.

18. Possibly the most famous individual among this group is the award-winning writer Arundhati Roy, who joined the struggle in the late 1990s and, according to

IRN's director, Patrick McCully, has greatly contributed to revitalize the involvement and interest of the Indian middle classes in the plight of Narmada populations (McCully's interview on WBEZ Radio Chicago, November 7, 2000).

19. The Narmada network is actively attempting to strengthen its links to this community as one strategy to confront setbacks that the mobilization has suffered since 1999. One example of this effort is the workshop "Resisting Unjust Development: Building Alliances to Support People's Struggles in India," originally scheduled to meet in San Francisco, October 12–14, 2001 and later postponed to 2002. The "workshop will explore ways in which (the Non Resident Indian) community can work together for social justice and sustainable development" (Letter by Juliette Majot and Malavika Vartak, International Rivers Network, September 7, 2001).

20. Japan's Overseas Economic Cooperation Fund (OECF) was one major source of funds for the SSP. During the 1990 International Narmada Symposium in Tokyo members of the Narmada network actively lobbied the Japanese government to withdraw support for the project. The symposium heightened Japanese public opinion's opposition to the project and, within a month of the meeting, Japan announced that it would no longer finance the SSP (Udall 1995).

21. Baviskar 1995.

22. Kothari 1995; Patkar 1995.

23. For instance, Suzanne Moxon, "All Aboard the Eco-bandwagon," *International Water Power & Dam Construction*, February 1998, 18–19, and Ramchandra Guha, "The Arun Shourie of the Left," *The Hindu Magazine*, November 26, 2000.

24. Fisher 1995 and Patkar 1995.

25. Patel and Metha 1995, and for a somewhat different account of events, see Patkar 1995.

26. Patkar's interview with WBEZ Radio.

27. For details on local strategies, see Baviskar 1995, Patkar 1995, and McCully 1996.

28. Repression reached such high levels that forty-three human rights, environmental, and indigenous organizations from sixteen countries set up a panel in 1993 to document abuses (Baviskar 1995, Udall 1995).

29. See note 20 above.

30. McCully 1996.

31. Patkar 1995.

32. Udall 1995, 216, citing the Morse Report 1992, xii.

33. See details of international lobby strategies in McCully 1996, Udall 1995, and Rich 1994.

34. Human rights abuses escalated from October 1992, to March 1993 (Udall 1995) and beyond (Sen 2000).

35. In 1991, the Congress Party obtained 37.3 percent of the popular vote and the BJP and allies 19.9 percent. In 1996, the BJP's percentage of votes had climbed to 24 and in the 1998 elections the party obtained 36.2 percent of votes, thus forming the government (Kesselman 2000, 274).

36. "The Wapenhans Report" or the Report of the Portfolio Management Task Force.

37. According to McCully (1996), "The independent Inspection Panel established in 1993 to assess violations of Bank policies is a direct result of the institution's humiliating experience with Sardar Sarovar" (p. 308).

38. *www.dams.org/about/debate*.

39. The document was eventually released to the public in response to a December 1994 order by India's supreme court.

40. Prajapati 1997.

41. *www.narmada.org/sardar-sarovar/sc.ruling*.

42. Kesselman et al. 2000, 269.

43. Mahesh Unydal, "Lawmakers Furious Over Court Ban on Dam," in Inter Press Service, March 9, 1997 *(www.oneworld.org/ips2/mar/india)*.

44. Prajapati 1997.

45. NBA, "The Narmada Struggle—The International Campaign After the World Bank Pull Out," October 1995 (mimeo).

46. Kothari 1995.

47. NBA e-mail update to U.S. environmental and human rights NGOs, July 1, 1995.

48. Kothari 1995.

49. Roy 1999, 9.

50. Baviskar 1995 and Sen 2000.

51. Roy 1999, 41.

52. See Introduction for a summary of the literature that discusses transnational advocacy networks' effectiveness.

53. Letter from Shripad Dharmadhikary, NBA activist, to Suzanne Moxon, writer for the *International Water Power and Dam Construction*, April 4, 1998.

54. Baviskar 1995, 222.

55. Fisher 1995, 23.

56. When discussing one of the Andolan's strategies, Patkar explains that "[a] basic principle was that the people's representatives should accompany the activists every time we met officials" (Patkar 1995, 158).

57. Baviskar (1995) puts the issue in the following terms: "It must be noted that (NBA's) critique of development has been formulated by the activists in the movement

and by supporters outside the valley; it is not the creation of people in the valley, both Adivasis and non-Adivasis, who understand the issue of displacement in a much more particularistic way" (p. 222). See also the critique made by Esteva and Prakash (1992).

Chapter 8. Conclusions

1. The reader should notice that this presents yet another aspect of International NGOs' political responsibility in transnational activism not fully explored by Jordan and Tuijl (2000).

2. Lipshutz and Conca 1993.

3. Which have been pursued, for instance, by Putnam 1988, and Keck and Sikkink 1998.

REFERENCES

Adams, William M. 1990. *Green Development—Environment and Sustainability in the Third World*. London and New York: Routledge.

Archibugi, F., P. Nijkamp, and F. Soeteman. 1989. "The Challenges of Sustainable Development." In *Economy and Ecology: Towards Sustainable Development*, ed. F. Archibugi, and P. Nijkamp. Dordrecht/Boston/London: Kluwer.

Aufderheide, Pat, and Bruce Rich. 1985. "Debacle in the Amazon," *Defenders* (Mar/Apr.).

Baviskar, Amita. 1995. *In the Belly of the River—Tribal Conflicts over Development in the Narmada Valley*. New Delhi, Bombay, Calcutta, and Madras: Oxford University Press.

Bebbington, Anthony. 1998. "Commentary to Seeking Common Ground in Ecuador," *Environment* 40, no. 5 (June).

Benavides, Margarida. 1996. "Amazon Indigenous Peoples—New Challenges for Political Participation and Sustainable Development," *Cultural Survival Quarterly* 20, no. 3 (Fall).

Binswanger, Hans. 1990. *Brazilian Policies that Encourage Deforestation in the Amazon*. Environmental Department Working Paper n. 16, The World Bank, Washington, D.C.

Blaikie, Piers. 1985. *The Political Economy of Soil Erosion in Developing Countries*. London and New York: Longman.

Boissevain, Jeremy. 1974. *Friends of Friends—Networks, Manipulators, and Coalitions*. Oxford: Basil Blackwell.

Brysk, Alison. 2000. *From Tribal Village to Global Village—Indian Rights and International Relations in Latin America*. Stanford: Stanford University Press.

Browder, John. 1987. "Brazil's Export Promotion Policy (1980–1984). Impacts on the Amazon's Industrial Wood Sector." *The Journal of Development Areas* 21 (April).

———. 1998. "World Bank Mid-Term Supervision Mission for the Program for the Support to Community Initiatives (PAIC), PLANAFLORO (loan 3444–BR), November 20–December 3, 1998." Mimeo, December 18, 1998.

Bunker, Stephen. 1985. *Underdeveloping the Amazon—Extraction, Unequal Exchange, and the Failure of the Modern State.* Chicago and London: University of Chicago Press.

Carvalho, Georgia et al. 2001. "Letter to the Editor: Sensitive development could protect Amazonia instead of destroying it," *Nature* 409:131.

Carvalho Filho, José, coord. 1987. "Avaliação Conjuntural do Polonoroeste (OS N. 1-1986/87-Relatório 1.5)." São Paulo: FIPE.

Castells, Manuel. 1996. *The Rise of the Network Society,* second edition. Malden, Mass.: Blackwell Publishers.

The Center for Economic and Social Rights (CESR), Oxfam America, and Rainforest Action Network (RAN). 1996. "Oil in the Rainforest—The Impact of and Responses to Texaco's Operations in the Ecuadorian Amazon." Information brochure, November.

Centro Ecumênico de Documentação e Informação (CEDI), and Coordenação Nacional de Geólogos (CONAGE). 1986. *Empresas de Mineração e Terras Indígenas na Amazônia.* São Paulo: CEDI and CONAGE.

Clark, Dana, Jonathan Fox, and Kay Treakle, eds. 2003 (forthcoming). *Demanding Accountability: Civil Social Claims and the World Bank Inspection Panel.* Lanham: Rowman and Littlefield.

Collins, Jennifer. 2000. "A Sense of Possibility—Ecuador's Indigenous Movement Takes Center Stage." *NACLA Report on the Americas* XXXIII, no. 5 (March/April).

Dalton, Russell. 1993. "The Environmental Movement in Western Europe." In *Environmental Politics in the International Arena—Movements, Parties, Organizations, and Policy,* ed. S. Kamieniecki. Albany: State University of New York Press.

Davis, Shelton. 1977. *Victims of the Miracle—Development and Indians of Brazil.* Cambridge: Cambridge University Press.

Dreifuss, René. 2000. "Strategic Perceptions and Frontier Policies in Brazil." In *Amazonia at the Crossroads—The Challenges of Sustainable Development,* ed. Anthony Hall. London: Institute of Latin American Studies.

Esteva, Gustavo, and Madhu Prakash. 1992. "Grassroots Resistance to Sustainable Development: Lessons from the Banks of the Narmada," *The Ecologist* 22, no. 2 (March/April).

Fearnside, Philip. 1989. *A Ocupação Humana de Rondônia—Impactos, Limites e Planejamento.* Programa Polonoroeste, Relatório de Pesquisa n. 5. SCT/PR CNPq.

Fisher, William. 1995. "Development and Resistance in the Narmada Valley." In *Toward Sustainable Development—Struggling over India's Narmada River,* ed. William Fisher. Armonk, N.Y., and London: M. E. Sharpe.

Fonseca, Francisco. 1981. "Projeto Carajás," *Ciências da Terra* 1 (November).

Forum das ONGs e Movimentos Sociais Que Atuam em Rondônia, and Friends of the Earth/Amigos da Terra Programa Amazônia. 1995. *Request for Inspection sub-*

mitted to the World Bank Inspection Panel on the Planafloro—Rondônia Natural Resources Management Project, Porto Velho, July 25.

Fox, Jonathan. 1992. *The Politics of Food in Mexico—State Power and Social Mobilization.* Ithaca and London: Cornell University Press.

———. 2000. "The World Bank Inspection Panel: Lessons from the First Five Years," *Global Governance* 6, no. 3 (July-September).

———. 2000b. "Civil Society and Political Accountability: Propositions for Discussion." Paper presented at "Institutions, Accountability and Democratic Governance in Latin America," Mimeo. The Helen Kellogg Institute for International Studies, May 8–9.

Fox, Jonathan, and L. David Brown, eds. *The Struggle for Accountability: The World Bank, NGOs, and Grassroots Movements.* Cambridge, Mass. and London: The MIT Press.

Freitas, Alencar, and Pedro Soares, org. 1994. *Aspectos Ambientais de Projetos Co-Financiados Pelo Banco Mundial.* Série IPEA 146. Brasília: IPEA.

Frey, Frederick. 1985. "The Problem of Actor Designation in Political Analysis," *Comparative Politics* 17, no. 2 (January).

Goodland, Robert. 1985. "Environmental Aspects of Amazonian Development Projects in Brazil." Office of Environmental and Scientific Affairs—Projects Policy Dept., The World Bank, Washington, D.C.

Hall, A. 1989. *Developing Amazonia—Deforestation and Social Conflict in Brazil's Carajas Programme.* Manchester and New York: Manchester University Press.

———, ed. 2000. "Introduction." In *Amazonia at the Crossroads—The Challenges of Sustainable Development.* London: Institute of Latin American Studies.

Haas, Peter. 1989. "Do Regimes Matter? Epistemic Communities and Mediterranean Pollution Control," *International Organization* 43, no. 3 (Summer).

Hecht, Suzanna. 1983. "Cattle Ranching in the Eastern Amazon: Environmental and Social Implications." In *The Dilemma of Amazonian Development,* ed. E. Moran. Boulder: Westview Press.

———. 1985. "Environment, Development, and Politics: Capital Accumulation and the Livestock Sector in Eastern Amazonia," *World Development* 13, no.6.

Hecht, Suzanna, and Alexander Cockburn. 1989. *The Fate of the Forest.* London and New York: Verso.

Heclo, Hugh. 1978. "Issue Networks and the Executive Establishment." In *The American Political System,* ed. Anthony King. Washington, D.C.: American Enterprise Institute for Public Policy Research.

Instituto Brasileiro de Análises Sociais e Econômicas (IBASE). 1985. *A Amazônia Legal: Políticas de Desenvolvimento e Seus Efeitos Econômicos e Sociais—Dossiê Amazônia.* Rio de Janeiro: IBASE.

———. 1990. *Política de Desenvolvimento Regional na Amazônia—20 Anos de SUDAM.* Rio de Janeiro: IBASE.

Instituto Brasileiro de Geografia e Estatística (IBGE). 1993. *Recursos Naturais e Meio Ambiente—Uma Visão do Brasil*. Rio de Janeiro: IBGE.

———. 1994. *Anuário Estatístico do Brasil—1993*. Rio de Janeiro: IBGE.

Instituto Nacional de Pesquisas Espaciais (INPE). 1992. *Deforestation in Brazilian Amazon*. São Jose dos Campos: INPE, May.

Inzet Association. 1999. "Civil Society Participation in a New EU-ACP Partnership." Mimeo. Report of a Workshop, Amsterdam, 11–12 January.

Jelin, Elizabeth, and Eric Hershberg. 1996. *Constructing Democracy—Human Rights, Citizenship, and Society in Latin America*. Boulder: Westview Press.

Jesus, Eli. 1998. "Relatório Final de Avaliação dos Projetos de Iniciativa Comunitária Implementados Pelo Planafloro em 1995 e 1996." Mimeo. Rio de Janeiro, November.

Jezic, Tamara. 2001. "Ecuador: The Campaign Against Texaco Oil." In *Advocacy for Social Justice—A Global Action and Reflection Guide*, ed. D. Cohen, R. Vega, and G. Watson. Bloomfield, Conn.: Kumarian Press.

Jochnick, Chris, and Paulina Garzon. 2001. " A Seat at the Table," *NACLA Report on the Americas* XXXIV, no. 4 (January/February).

Jordan, Lisa, and Peter Van Tuijl. 2000. "Political Responsibility in Transnational NGO Advocacy," *World Development* 28, no. 12.

Kane, Joe. 1996. *Savages*. New York: Vintage Books.

Keck, Margaret. 1993. "The Western Amazon: Planafloro in Rondônia, Brazil." Paper presented at the first meeting of authors for the project The Struggle for Accountability—The World Bank, NGOs, and Grassroots Movements, organized by Jonathan Fox and David L. Brown, Department of Political Science, Massachusetts Institute of Technology.

———. 1995. "Social Equity and Environmental Politics in Brazil: Lessons from the Rubber-Tappers of Acre," *Comparative Politics* 27, no. 4 (July).

———. 1998. "Planafloro in Rondônia: The Limits of Leverage." In *The Struggle for Accountability—The World Bank, NGOs, and Grassroots Movements*, ed. Jonathan Fox and David L. Brown. Cambridge, Mass. and London: The MIT Press.

Keck, Margaret, and Kathryn Sikkink. 1998. *Activists Beyond Borders—Advocacy Networks in International Politics*. Ithaca and London: Cornell University Press.

Kesselman, Mark, Joel Krieger, and William Joseph, eds. 2000. *Introduction to Comparative Politics*, second edition. Boston: Houghton Mifflin.

Kennedy Jr., Robert, and S. Jacob Scen. n/d. "Response to 'With Spears From All Sides' by Joe Kane, *The New Yorker*, September 27, 1993, pp. 54–79," New York: Natural Resources Defense Council (NRDC).

Kimerling, Judith. 1991. *Amazon Crude*. New York: Natural Resources Defense Council.

Kolk, Ans. 1998. "From Conflict to Cooperation: International Policies to Protect the Brazilian Amazon," *World Development* 26, no. 8: 1481–1493.

Kothari, Smitu. 1995. "Damming the Narmada and the Politics of Development." In *Toward Sustainable Development—Struggling over India's Narmada River*, ed. William Fisher. Armonk, N.Y., and London: M.E. Sharpe.

Lélé, Sharachchandra. 1991. "Sustainable Development: A Critical Review," *World Development* 19, no. 6.

LePreste, Phillip. 1989. *The World Bank and the Environmental Challenge*. Selinsgrove, Penn.: Susquehanna University Press.

Leonel, Mauro. 1991. "Colonos Contra Amazonidas no Polonoroeste: Uma Advertência às Políticas Públicas." In *Amazônia—A Fronteira Agrícola 20 Anos Depois*, ed. Philippe Lena and Adelia Oliveira. Belém: Museu Paraense Emílio Goeldi.

Lipschutz, Ronnie, with Judith Mayer. 1996. *Global Civil Society and Global Environmental Governance—The Politics of Nature from Place to Planet*. Albany: State University of New York Press.

———. 1997. "From Place to Planet: Local Knowledge and Global Environmental Governance," *Global Governance* 3.

Lipschutz, Ronnie, and Kenneth Conca, eds. 1993. *The State and Social Power in Global Environmental Politics*. New York: Columbia University Press.

Mahar, Dennis. 1989. *Government Policies and Deforestation in Brazil's Amazon Region*. Washington, D.C: The World Bank.

Maybury-Lewis, David et al. 1981. *In the Path of Polonoroeste: Endangered Peoples of Western Brazil*. Cambridge: Cultural Survival.

McCormick, John. 1989. *Reclaiming Paradise—The Global Environmental Movement*. Bloomington and Indianapolis: Indiana University Press.

———. 1993. "International Nongovernmental Organizations: Prospects for a Global Environmental Movement." In *Environmental Politics in the International Arena—Movements, Parties, Organizations, and Policy*, ed. S. Kamienecki. Albany: State University of New York Press.

McCreary, Scott. 1995. "Independent Fact-Finding as a Catalyst for Cross-Cultural Dialogue—Assessing Impacts of Oil and Gas Development in Ecuador's Oriente Region," *Cultural Survival Quarterly* 19, issue 3 (Fall).

McCully, Patrick. 1996. *Silenced Rivers—The Ecology and Politics of Large Dams*. London: Zed Books.

Mendez, Sixto, Jennifer Parnell, and Robert Wasserstrom. 1998. "Seeking Common Ground—Petroleum and Indigenous Peoples in Ecuador's Amazon," *Environment* 40, no. 5 (June).

Milbrath, Lester. 1993. "The World Is Relearning Its Story about How the World Works." In *Environmental Politics in the International Arena—Movements, Parties, Organizations, and Policy*, ed. S. Kamienecki. Albany: State University of New York Press.

Migdal, Joel, A. Kohli, and V. Shue, eds. 1994. *State Power and Social Forces—Domination and Transformation in the Third World.* New York: Cambridge University Press.

Millikan, Brent. 1992. "Tropical Deforestation, Land Degradation, and Society—Lessons from Rondônia, Brazil," *Latin American Perspectives* 72, vol. 19, no. 1 (Winter).

———. 1995. "A Campanha do Forum-Rondônia e a Situação Atual de Implementação do Planafloro." Rio de Janeiro: *FASE* (newsletter).

———. 1998. "Planafloro: Modelo de Projeto Participativo." In *Bancos Multilaterais e Desenvolvimento Participativo no Brasil: Dilemas e Desafios,* ed. Jean-Pierre Leroy and Maria Clara Soares. Rio de Janeiro: FASE/IBASE.

———. 2001. "O Painel de Inspeção do Banco Mundial e o Pedido de Investigação sobre o Planafloro." In *Banco Mundial: Participação, Transparência e Responsabilidade: A Experiência Brasileira com o Painel de Inspeção,* ed. Aurélio Vianna Jr. et al. Brasília: Rede Brasil.

Mindlin, Betty. 1988. "Índios e Grandes Projetos Econômicos—Um Exemplo: O Programa Polonoroeste (Mato Grosso e Rondônia, 1982–87)." Mimeo. São Paulo: Instituto de Antropologia e Meio Ambiente.

Mitchell, Robert C. 1990. "Public Opinion and the Green Lobby: Poised for the 1990s?" In *Environmental Policy in the 1990s,* ed. Norman Vig and Michael Kraft. Washington, D.C.: CQ Press.

Mohan, Giles, and Kristian Tokke. 2000. "Participatory Development and Empowerment: The Dangers of Localism," *Third World Quarterly* 21, no. 2.

Morse, Bradford, and Thomas Berger. 1995. "Findings and Recommendations of the Independent Review." In *Toward Sustainable Development—Struggling over India's Narmada River,* ed. William Fisher. Armonk, N.Y., and London: M.E. Sharpe.

Moore, Sara, and Maria Carmem Lemos. 1999. "Indigenous Policy in Brazil: The Development of Decree 1775 and the Proposed Raposa/Serra do Sol Reserve, Roraima, Brazil," *Human Rights Quarterly* 21.

Moran, Emilio. 1981. *Developing the Amazon.* Bloomington: Indiana University Press.

Norgaard, R. B. 1984. "Co-evolutionary Development Potential," *Land Economics* 60, no. 6.

O'Brien, Robert, A. Goetz, J. Scholte, and M. Williams. 2000. *Contesting Global Governance—Multilateral Economic Institutions and Global Social Movements.* Cambridge: Cambridge University Press.

O'Donnell, Guillermo, and Phillip Schmitter. 1991. *Transitions from Authoritarian Rule.* Baltimore: John Hopkins University Press.

Oliveira, Luiz Rodrigues. 1998. "Considerações Sobre a Participação da Sociedade Civil no Programa de Apoio a Iniciativas Comunitárias—PAIC/Planafloro," *O Futuro da Amazônia em Questão,* Cadernos de Proposta, no. 77 (June/August). Rio de Janeiro: FASE.

Oliveira Filho, J. Pacheco. 1990. "Frontier Security and the New Indigenism: Nature and Origins of the Calha Norte Project." In *The Future of Amazonia—Destruction or Sustainable Development?*, ed. D. Goodman and A. Hall. New York: St. Martin's Press.

O'Riordan, Thomas. 1985. "Future Directions in Environmental Policy," *Journal of Environmental and Planning* 17.

Ortiz-T, Pablo. 1997. "Ecuador's Indigenous People: 'We Seek True Participation,'" *Cultural Survival Quarterly* 21, issue 2 (Summer).

Pacari, Nina. 1996. "Ecuador—Taking On the Neoliberal Agenda," *NACLA Report on the Americas* XXIX, no. 5 (March/April).

Pádua, José A. 1990. *Ecologia e Política no Brasil*. Rio de Janeiro: IUPERJ.

Patel, Anil. 1995. "What Do the Narmada Valley Tribals Want?" In *Toward Sustainable Development—Struggling over India's Narmada River*, ed. William Fisher. Armonk, N.Y., and London: M.E. Sharpe.

Patel, Anil, and Ambrish Mehta. 1995. "The Independent Review: Was It a Search for Truth?" In *Toward Sustainable Development—Struggling over India's Narmada River*, ed. William Fisher. Armonk, N.Y., and London: M.E. Sharpe.

Peterson, Matthew. 1999. "Interpreting Trends in Global Environmental Governance," *International Affairs* 75, no. 4.

Patkar, Medha. 1995. "The Struggle for Participation and Justice: A Historical Narrative (Patkar in conversation with Smitu Kothari)." In *Toward Sustainable Development—Struggling over India's Narmada River*, ed. William Fisher. Armonk, N.Y., and London: M.E. Sharpe.

Pearce, David, E. Barbier, and A. Markandya. 1990. *Sustainable Development—Economics and Environment in the Third World*. Surrey, U.K., and Northampton, Mass.: London Environmental Economics Centre, Edward Elgar publishing Ltd.

Pearce, David, and N. Myers. 1990. "Values and the Environment of Amazonia." In *The Future of Amazonia—Destruction or Sustainable Development?*, D. Goodman and A. Hall. New York: St. Martin Press.

Pearce, David, and John Warford. 1993. *World Without End—Economics, Environment, and Sustainable Development*. Washington, D.C., and New York: The World Bank and Oxford University Press.

Pinheiro, Nilde, and Maria A. Leão. 1989. "Avaliação 'ex post' do Programa Integrado de Desenvolvimento do Noroeste do Brasil—Polonoroeste." Brasília: C.A.A./SEPES/SEPLAN-PR.

Prajapati, Rohit. 1997. "Narmada, the Judiciary, and Parliament," *Economic and Political Weekly*, April 5, Mumbay.

Pretty, Jules, and Hugh Ward. 2001. "Social Capital and the Environment," *World Development* 29, no. 2.

Price, David. 1980. "The Brazilian Capability for Protecting the Native Population in the Guapore Valley from the Effects of Project Polonoroeste." A Report to the World Bank.

———. 1989. *Before the Bulldozer—The Nambikwara Indians and the World Bank*. Cabin John, Md./ Washington, D.C.: Seven Lock Press.

Princen, Thomas, and Mattihas Finger. 1994. *Environmental NGOs in World Politics—Linking the Local and the Global*. London and New York: Routledge.

Princen, Thomas. 1994. "NGOs: Creating a Niche in Environmental Diplomacy." In *Environmental NGOs in World Politics—Linking the Local and the Global*. London and New York: Routledge.

Putnam, Robert. 1988. "Diplomacy and Domestic Politics: The Logic of Two-Level Games," *International Organization* 42, no. 3 (Summer).

Redclift, Michael. 1987. *Sustainable Development—Exploring the Contradictions*. London and New York: Methuen & Co.

Rich, Bruce. 1985. "The Multilateral Development Banks, Environmental Policy, and the United States," *Ecology Law Quarterly* 12, no.4.

———. 1987. "Environmental Management and Multilateral Development Banks," *Cultural Survival Quarterly* 10, no. 1.

———. 1989. "The 'Greening' of the Development Banks: Rhetoric and Reality," *The Ecologist* 19, no.2.

———. 1990. "The Emperor's New Clothes: The World Bank and Environmental Reform," *World Policy Journal* (Spring).

———. 1994. *Mortgaging the Earth—The World Bank, Environmental Impoverishment, and the Crisis of Development*. Boston: Beacon Press.

Rodrigues, Maria, 2000. "Environmental Protection Issue Networks in Amazonia," *Latin American Research Review* 35, no. 3.

Ros Filho, Luis Carlos. 1994. *Financiamentos Para o Meio Ambiente*. Brasília, D.F.: Instituto de Estudos Amazônicos e Ambientais, IEA.

Rosenau, James 1993. "Environmental Challenges in a Turbulent World." In *The State and Social Power in Global Environmental Politics*, ed. R. Lipschutz and K. Conca. New York: Columbia University Press.

Roy, Arundhaty. 1999. *The Cost of Living*. New York: The Modern Library.

Sawyer, Suzana. 1996. "Indigenous Initiatives and Petroleum Politics in the Ecuadorian Amazon," *Cultural Survival Quarterly* 20, issue 1 (Spring).

———. 1998. "Commentary to Seeking Common Ground in Ecuador," *Environment* 40, no. 5 (June).

Schmidt, James. 1997. "Civility, Enlightenment, and Society." Presented at the UNESCO conference on "Future Ethics" held at the Institute for Cultural Pluralism, Candido Mendes University, Rio de Janeiro, July 2–4, 1997.

Schmink, Marianne. 1992. "Amazonian Resistance Movements and the International Alliance." In *Ecological Disorder in Amazonia: Social Aspects*, ed. Leszek Kosinski. Rio de Janeiro: Unesco/ISSC/Educam.

Schartzman, Stephen. 1988. "Desenvolvimento, Meio Ambiente e Póvos Indígenas," *Tempo e Presença* 231 (May).

Selverston, Melina. 1997. "The Politics of Identity Reconstruction: Indians and Democracy in Ecuador." In *The New Politics of Inequality in Latin America—Rethinking Participation and Representation*, ed. Douglas Chalmers. Oxford: Oxford University Press.

Selverston-Scher, Melina. 2000. "Building International Civil Society: Lessons from the Amazon Coalition." Paper presented at the Conference "Human Rights and Globalization: When Transnational Civil Society Networks Hit the Ground," sponsored by the Center for Global, International, and Regional Studies at the University of California, Santa Cruz, December 1–2, 2000.

———. 2001. *Ethnopolitics in Ecuador: Indigenous Rights and the Strengthening of Democracy*. Miami: North-South Center Press.

Secretaria de Planejamento (SEPLAN)/Planafloro. 1997. Programa de Apoio à Iniciativa Comunitária—PAIC: Instruções Técnicas, Porto Velho, Rondônia (Manual de Operações, Caderno 1).

Shiva, Vandana. 1991a. *Ecology and the Politics of Survival*. New Delhi/Newbury Park/London: United Nations University Press, Sage Publications.

———. 1991b. *The Violence of the Green Revolution—Third World Agriculture, Ecology, and Politics*. London: Zed Books, Ltd.

Sikkink, Kathryn. 1993. "Human Rights, Principled Issue-Networks, and Sovereignty in Latin America," *International Organization* 47, no. 3 (Summer).

Smeraldi, Roberto, and Brent Millikan. 1997. "Planafloro—Um Ano Depois—Análise Crítica da Implementação do Plano Agropecuário e Florestal de Rondônia um Ano Após o Acordo Para a sua Reformulação." São Paulo and Porto Velho: Amigos da Terra Internacional—Programa Amazônia and Oxfam.

Tidwell, Mike. 1996. *Amazon Stranger*. New York: Lyons and Bwoford, Publishers.

Treakle, Kay. 1998. "Ecuador: Structural Adjustment and Indigenous and Environmentalist Resistance." In *The Struggle for Accountability: The World Bank, NGOs, and Grassroots Movements*, ed. J. Fox and D. Brown. Cambridge and London: The MIT Press.

Treece, Dave. 1987. *Bound in Misery and Iron—The Impact of the Grande Carajas Programm on the Indians in Brazil*. London: Survival International.

Udall, Lori. 1995. "The International Narmada Campaign: A Case of Sustained Advocacy." In *Toward Sustainable Development—Struggling over India's Narmada River*, ed. William Fisher. Armonk, N.Y., and London: M.E. Sharpe.

———. 1998. "The World Bank and Public Accountability: Has Anything Changed?" In *The Struggle for Accountability: The World Bank, NGOs, and Grassroots Movements*, ed. J. Fox and D. Brown. Cambridge and London: The MIT Press.

Vakil, A. C. 1997. "Confronting the Classification Problem: Toward a Taxonomy of NGOs," *World Development* 25, no. 2.

Wapner, Paul. 1996. *Environmental Activism and World Civic Politics.* Albany: State University of New York Press.

The World Bank. 1981. Brazil—Integrated Development of the Northwest Frontier. Washington, D.C.: The World Bank.

———. 1991. "Project Completion Report Brazil—Northwest Region Integrated Program—Phase I Agricultural Development and Environmental Protection Project (Loan 2060–BR)." Washington, D.C.: The World Bank.

———. 1992a. Staff Appraisal Report—Brazil Rondônia Natural Resources Management Project (February 27). Washington, D.C.: The World Bank.

———. 1992b. World Bank Approaches to the Environment in Brazil: A Review of Selected Projects—Volume V: The Carajas Iron Ore Program, Operations Evaluation Department, Washington, D.C.

———. 1992c. World Bank Approaches to the Environment in Brazil: A Review of Selected Projects—Volume V: The POLONOROESTE Program, Operations Evaluation Department, Washington, D.C.

World Commission on Environment and Development (WCED). 1987. *Our Common Future.* New York: Oxford University Press.

Young, Oran. 1972. "The Actors in World Politics." In *The Analysis of International Politics,* ed. James Rosenau et al. New York: Free Press.

INDEX

Acción Ecológica, 97, 101, 102, 106, 107, 109, 112
Achuar Federation, 110
Acre state, 52
Ação Ecológica para o Vale do Guaporé (Ecological Action for the Guaporé Valley—Ecoporé), 69, 82
Adivasi, 116, 119, 129
American Anthropological Association, 34
Amazanga Institute, 108, 111
Amazon Coalition, 97, 99, 100–102, 103, 107, 112, 144
Amazon Crude, 95
Amazonia por la Vida (Amazon for Life), campaign, 97, 98, 102
Amazonia, 12, 19; development policies, 23, 84; environmental policies, 26; forest, 52; Sistema de Vigilância da (SIVAM), 30; Superintendência para o Desenvolvimento da (SUDAM), 23
Amerindians, 24, 30, 34, 39, 40, 44, 46, 47, 85; Cofan, 96, 113; Ecuador, 96, 97–99, 100; Huaorani, 95; Karitiana, 58, 64; Nambikwara, 39; Quichua, 96; reserves, 25, 27, 29, 54, 62, 64, 74, 161n. 62; Secoya, 96; Siona, 96; Special Project, 40, 41; Yanomami, 30
Amte, Baba, 118
Anti-oil network, 93, 97, 110, 140; cleavages 99–101; resources 101–102

Arch-Vahini, 116, 117
ARCO Oriente 95, 96, 107, 108, 109, 110
Association for Indigenous Development (ASODIRA), 108, 109
Atlantic Ritchfield, 95
Avança Brasil, 31

Ballen, Durán, 99
Baviskar, Amita, 131
Belém-Brasília Highway, 21, 22
Berlim wall, 5
Better World Society, 52
Bharatiya Janata Party (BJP), 124, 127
Blackwelder, Brent, 36
Bolivia, 20
Bonifaz, Cristobal, 103, 105
Borja, Rodrigo, 98
Bramble, Barbara, 36
Brazil, 12, 15, 40, 73; Congress, 38; democratization, 25, 90; Federal Constitution, 1988, 25; government, 24, 25, 26, 29, 45, 72, 74, military regime, 20, 30, 35; Secretaria de Assuntos Estratégicos (Secretariat of Strategic Affairs, SAE), 30
Brazilian Indian Agency. *See* Fundação Nacional do Indio (FUNAI)
Brazilian Network on Multilateral Development Banks. *See* Rede Brasil

Caldwell, Adrian, 37
Calha Norte, 30, 31

192

Index

Carajás, 19, 23
Cardoso, Fernando Henrique, 31
Catholic Church, 97
Center for Economic and Social Rights (CESR), 97, 103, 112
Center for International Environmental Law (CIEL), 50, 71
Centro de Derechos Economicos y Sociales (CDES). *See* Center for Economic and Social Rights (CESR)
China, 73
Coalition for Amazonian Peoples and their Environment. *See* Amazon Coalition
Coca-Orellana, 105
Collor de Mello, Fernando, 27, 53
Colombia, 30
Comite de los Demandantes contra la Texaco (Committee of the Plaintiffs against Texaco), 105
Comissão Pastoral da Terra (Pastoral Land Commission, CPT), 51
Conable, Barber, 52
Conoco, 95, 101
Confederation of Indian Nations of Ecuadorian Amazon (CONFENIAE), 96, 100, 104, 109, 112
Conselho Indígena Missionário (Indigenous Peoples' Missionary Council, CIMI), 51, 62, 64
Conselho Nacional dos Seringueiros (National Council of Rubber Tappers, CNS), 50, 52, 55
Coordenação da União das Nações e Póvos Indígenas de Rondônia (Coordination of the Union of Indigenous Peoples and Nations of Rondônia (CUNPIR), 69, 83, 85, 89
Coordinating Body of Indigenous Organizations of the Amazon Basin (COICA), 101
Corporación de la Defensa de la Vida (CORDAVI), 101, 112
Cuba, 20
Cuiabá-Porto Velho Highway, 22, 24

Damien Foundation, 62
Decade of Destruction, The, 37
Deforestation, 24, 26, 40, 46
Demonstrative projects, 27
Departamento Nacional de Prospecção Mineral (National Department of Mineral Research, DNPM), 46
Dutch Organization for International Development (NOVIB), 71

Ecuador; agricultural development law, 98; Amazon region, 94; Constitution, 110; environmental groups, 99; economic policies, 94; government, 98, 106, 111, 173n. 33; military, 99; Ministry of Mines and Energy, 109
Environmental Defense (EDF). *See* Environmental Defense Fund (EDF)
Environmental Defense Fund (EDF), 7, 34, 36, 43, 50, 53, 55, 56, 118, 122
Environmental Impact Assessments (EIA), 58
Environmental Policy Institute, 36, 118
Environmentally Sustainable Development, Amazonia, 31; concept, 9–10, 52, 91, 141; literature, 10; Planafloro, 54, 68, 77
Extractive reserves, 27, 54, 63, 74, 83
Exxon Valdez, 95

Federação dos Trabalhadores Agrícolas de Rondônia (Federation of Agricultural Workers of Rondônia, FETAGRO), 62, 68, 82, 83, 89, 90
Fisher, William, 131
Forum de ONGs e Movimentos Sociais de Rondônia (Forum of NGOs and Social Movements of Rondônia). *See* Rondônia Forum
France, 73
Frente de Defensa de la Amazonia (Amazon Defense Front), 104–106, 111, 112

Friends of the Earth (FoE), 1, 2, 50, 71, 118
Fórum Brasil de ONGs e Movimentos Sociais para Meio Ambiente e Desenvolvimento (Brazilian National Forum of NGOs and Social Movements on the Environment and Development), 90
Fundação Instituto de Pesquisas Econômicas (Institute of Economic Research Foundation, FIPE), 40, 41, 44
Fundação Nacional do Indio (FUNAI), 25, 29, 34, 41, 45, 57, 85
Fundo Nacional para o Meio Ambiente (National Fund for the Environment, FNMA), 26

Garrison, John, 76
Geographic Information System (GIS), 109
Global environmental governance, 1, 2
Globalization, 8
Goldman Award, 118
Grito da Terra, 82
Gujarat state, 116, 123, 126; repression, 124

Hunter, David, 73

India, 73, 121; Congress party, 127; government, 120, 123, 125, 129; parliament, 126; police, 124; Supreme Court, 125, 126
Independent Federation of the Shuar People of Ecuador (FIPSE). *See* Shuar Federation
Indigenous peoples. *See* Amerindians
Instituto Brasileiro de Desenvolvimento Florestal (Brazilian Institute for Forestry Development, IBDF), 25
Instituto Brasileiro de Meio Ambiente e Recursos Naturais Renováveis (Brazilian Institute for the Environment and Renewable Natural Resources, IBAMA), 28, 29

Instituto de Antropologia e Meio Ambiente (Institute of Anthropology and the Environment, IAMÁ), 50, 52
Instituto de Estudos Amazônicos e Ambientais (Institute for Amazonian and Environmental Studies, IEA), 7, 50, 52, 53, 55, 56
Instituto de Pré-História, Antropologia, e Ecologia (Institute of Pre-History, Anthropology, and Ecology, IPHAE), 69
Instituto Nacional de Colonização e Reforma Agrária (National Institute for Colonization and Agrarian Reform, INCRA), 29, 58, 59, 60, 62, 74, 164n. 30
Inter-American Development Bank (IDB), 21, 26, 98
International Development Association (IDA), 37
International Governmental Organizations (IGOs), 2
International Rivers Network (IRN), 118

Japan, 122, 128
Jezic, Tamara, 100

Kasten, Robert, 37, 38
Keck, Margaret, 2
Kimerling, Judith, 95, 97

Lévi-Strauss, Claude, 39
Lokayan network, 117
Lutzenberger, José, 35, 37, 53

Madeira river, 54
Madhya Pradesh state, 116, 126
Maharashtra state, 116, 117
Mahuad, Jamil, 99, 103
McCormick, John, 7
MDB campaign, 34, 35, 36, 37, 40, 44, 47, 50, 51, 52, 55, 57, 58, 59, 117, 119, 137
Mendes, Chico, 26, 50, 52

194　　　　　　　　　Index

Morse, Bradford, 123; report, 123, 125
Morona-Santiago, province, 109
Movimento dos Trabalhadores Rurais Sem Terra (Landless Peoples' Movement, MST), 51, 69, 82
Multilateral Development Banks campaign. *See* MDB campaign

Narmada; Action Committee, 122; control authority, 126; network, 115, 118, 122, 127, 128, 129, 140; river 115, valley, 115, 116; Water Dispute Tribunal, 126
Narmada Bachao Andolan (NBA), 117, 118, 119, 120, 122, 123, 125, 129, 130, 131, 132, 142, 143
National Confederation of Indigenous Nationalities of Ecuador (CONAIE), 96, 98, 99, 100, 108, 109, 112
National Wildlife Federation (NWF), 34, 36, 43, 55, 118
Natural Capital Stock, 11
Natural Resources Defense Council (NRDC), 95, 96, 101
New York, 104
Nimar valley, 117, 121, 129
Nogaard, R. B., 11
Non-Governmental Organizations (NGOs), 2, 135; concept, 6–7, 117, 137; international, 3, 34, 44, 50, 69, 90, 96, 101
Northwest Brazil Integrated Development Program. *See* Polonoroeste
Nossa Natureza, 26, 27, 29

Operação Amazônia, 20, 21
Organização dos Seringueiros de Rondônia (Organization of Rondonian Rubber Tappers—OSR), 1, 7, 53, 61, 63, 68, 83, 89
Organization of Indigenous Peoples of Pastaza (OPIP), 96, 107, 108, 109, 113
Oxfam, 1, 50, 55, 56, 62, 71, 82; America, 97, 108, 109

Pachakutik, 99
Paiva, Marcos Caramuru, 73
Panamerican Highway, 98
Partido dos Trabalhadores, (The Workers' Party, PT), 74, 79
Pastaza region, 107, 108, 109
Patidars, 116, 118, 119
Patkar, Medha, 117, 122, 123, 125, 131
Peru, 10; Iquitos, 101
Petroecuador, 94
Planafloro, 28, 29, 49, 54; Deliberative Council, 56; evaluation seminar, 75, 78; implementation, 63, 65, 74, 83; inspection panel, 68, 71–75, 87–88; interruption, 59, 60, 61; negotiations, 50, 51; zoning, 28, 83
Plan Colombia, 31
Plano de Integração Nacional (National Integration Plan, PIN), 22, 23
Polamazonia, 23
Polonoroeste, 24, 25, 28, 34, 35, 36, 37, 39, 40, 44, 45, 46
Porto Velho, 55
Programa de Apoio a Iniciativas Comunitárias (Program for the Support of Community Initiatives, PAIC), 79–88, 139, 141
Programa Nacional para o Meio Ambiente (National Program for the Environment (PNMA), 27, 28, 47
Programa Piloto para a Conservação de Florestas Tropicais, 27, 47
Protocolo de Entendimento, 56, 163n. 18

Quito, 102, 105

Rainforest Action Network (RAN), 62, 97, 108, 112
Raupp, Walter, 74, 86–87
Raytheon, 30
Red de Monitoreo Ambiental, 104, 111
Rede Brazil, 90
Rich, Bruce, 36
Right Livelihood Award, 118
Rondônia Natural Resources Management Project. *See* Planafloro

Rondônia state, 1, 12, 43, 47; deforestation, 23; development policies, 19, 45; Federação das Indústrias (FIERO), 75; government, 28, 49, 51, 52, 55, 68, 72, 75, 78; logging industry, 46, 64; zoning plan, 164n. 24

Rondônia Forum, 56, 58, 60, 61, 77, 130, 131, 132, 138, 142; crisis, 59, 64, 69, 72; identity, 65; political leverage, 62, 82, 89

Rondônia network, 12–13, 29, 49, 45, 76, 130, 137; effects, 47, 48, 52, 64; legitimacy crisis, 50, 60, 63, 69, 138; membership, 35; resources, 53, 64; strategies, 36, 38, 40, 55, tensions, 44, 45, 59; weakness, 88

Roy, Arundhaty, 127, 128

Sangath movement, 116
Sardar Sarovar Project (SSP), 115, 116, 115, 118, 119, 121, 122, 129; independent review, 123
Sarney, José, 26, 29
Satyagraha, 122, 124
Schwartzman, Stephen, 53, 60
Selverston-Scherr, Melina, 100, 101
Seventh Generation Fund, 109
Shramik Sanghatan, 117
Shuar Federation, 110
Sikkink, Kathryn, 2
Sindicato dos Trabalhadores Rurais (Rural Workers' Union), 52
Sucumbios, 105

Taloda forest, 120
Tata Institute of Social Science, 117
Teixeira, Emerson, 79
Texaco, 95, 97, 102, 103, 106; lawsuit, 103, 105, 106, 107

Tocqueville, Alexis de, 7
Trans-Amazonia highway, 22
Transnational advocacy networks, 2, 3, 5, 90; effectiveness, 135–136; methodology, 10, 135; strategies, 14
Transnational environmental advocacy networks, 2, 3, 67, 91; effectiveness, 4; membership, 13
Tropical forests, 44, 84

União Nacional Indígena (National Indigenous Union, UNI), 55, 58, 64
United Kingdom, 73
United Nations Conference on Environment and Development (UNCED), 10, 16, 27, 28, 52, 138
United Nations Environmental Program (UNEP), 52
United States, Congress, 26, 34, 35, 43, 52, 122; district court, 103
University of California, Berkeley, 108

Villano Assembly, 107, 108

Washington, D.C., 1, 52
West German Bundestag, 38
World Bank, 1, 4, 21, 24, 25, 27, 33, 35, 37, 38, 40, 45, 47, 49, 57, 58, 60, 61, 68, 75, 78, 111; environmental staff, 43; inspection panel, 72–73, 128; Narmada Campaign, 120, 122, 124, 128
World Commission on Dams, 125, 128, 148
World Commission on Environment and Development (WCED), 10
World Wildlife Fund for Nature, 56, 62, 82

Yanza, Luis, 106

www.ingramcontent.com/pod-product-compliance
Ingram Content Group UK Ltd.
Pitfield, Milton Keynes, MK11 3LW, UK
UKHW041919140426
5217IPUK00013B/221